Digital Transformation Shaping the Subconscious Minds of Organizations

Werner Leodolter

Digital Transformation Shaping the Subconscious Minds of Organizations

Innovative Organizations and Hybrid Intelligences

 Springer

Werner Leodolter
Center of Entrepreneurship and Applied
   Business Studies
University of Graz
Graz
Austria

ISBN 978-3-319-53617-0   ISBN 978-3-319-53618-7   (eBook)
DOI 10.1007/978-3-319-53618-7

Library of Congress Control Number: 2017945715

Printed on acid-free paper

This Springer imprint is published by Springer Nature
The registered company is Springer International Publishing AG
The registered company address is: Gewerbestrasse 11, 6330 Cham, Switzerland

To Konstantin, my grandson

Foreword

Evolution has equipped humans with specific abilities that reside in our brains and which give us dominion over land, sea, and earth including all other species. With the intelligence provided by this brain, mankind has developed speech, technology, and complex social organizations with each generation building upon the achievements of prior generations.

In all those developments also subconscious processes play important roles—especially when it comes to perceiving, recognizing, evaluating, deciding, or acting. This also applies when humans make decisions in organizations—the efficient "thinking fast" with all its biases and prejudices evoking the laborious "thinking slow" when it comes to complex issues.

Organizations are complex social structures where humans are essential elements and which are controlled by them. In recent years, new technologies have not just changed many parameters in organizations, but also influenced the way how decisions in organizations are made. In this context also Artificial Intelligence, which was developed by re-engineering the human brain, plays an important role. Combined with the human intelligence in organizations, this phenomenon could be considered as hybrid intelligence.

In this book, Werner Leodolter offers a framework for understanding changes in organizations when using new information and communication technologies. This framework with the metaphor of the "subconscious mind of organizations" supports managers to cope with inevitable developments in a dynamic environment—the digital transformation of their organization. Short, lucid stories support the conceptual ideas and form narrations about the future of manufacturing, health care, and retail.

Hybrid intelligence in combination with a subconscious mind could build the foundation of a future-oriented understanding of organizations. The more complex and international organizations are, the more dynamic the development of information technology is, the more importance this understanding will gain.

Thomas Foscht, Ph.D.
Professor of Marketing, Dean of the School of Business,
Economics and Social Studies, Karl-Franzens-University of Graz,
Graz, Austria

Preface

When I first learned to drive, I was fully and consciously concentrated on driving, on every gear change, use of the indicator, etc. After a relatively short period of time and frequent driving it started to become automatic and I could turn my attention to other things—talking with my passenger, listening to the radio, etc. A few years ago, when I started to commute to work with a more top of the range car, I suddenly had all sorts of options available: automatic drive, car phone, voice recognition, navigation system, parking sensors, proximity warning, etc.

The route to work was nearly all motorway. I began to use the time in the car more productively: teleconferences, dictations, etc. My beloved wife and my friends have always warned me of inattentiveness and accidents. Thank God that never happened, even if there were a few near accidents which would probably not had happened when driving at full attention. When I hear about automatic driving on motorways with self-steering vehicles—the automatic parking is already state of the art—it means that I will indeed soon be able to read and write while "driving" the car. Having emails read for you already works. But will I also be able to react properly in specific situations? Will I even recognize them? Will our children and grandchildren be at all capable of spatial/geographic orientation without GPS and a navigation system, or mistakenly enter the wrong way not even realizing that they are traveling in the completely wrong direction?

These banal observations on the one hand show the learning ability of people and the skills of the human brain. On the other hand, they show the ongoing expansion of the action spectrum of people with new tools—a fundamental development of mankind—starting with the first tools and hunting weapons, on to the invention of the wheel and further to the tools that the information and communication technologies provide at increasingly rapid succession.

Virtual reality and the expanding possibilities of perception through sensors, cameras, smart glasses, etc. accompany us as human beings and individuals, into a new era and on to a new stage of development in which our perception and action spectrum are significantly expanded.

Figuratively speaking, our arms are constantly expanding—the first tools primarily served as "extension" of our arms and their effectiveness—always becoming longer, swifter, more efficient, and more powerful. Organizations as "purposeful associations of people" massively change with the "length of our arms," the might of our tools, and the extended perception and actions. The design of organizations has to be thought in new ways. We should focus In particular on the decision making processes. Their designers, members, managers, and stakeholders are challenged to use these opportunities in an evolutionary way and—partly—to take advantage of revolutionary developments. Regarding our organizations, these developments in the field of "tools" for perception and actions are likely to be seen as "disruptive technologies and innovations." So we are challenged to "rethink" our organizations or at least think about whether they are still relevant and appropriate. This is facilitated by new perspectives and approaches.

The human brain may well be regarded as the crown of evolution and the human being as the crown of creation. A major part of the capacity of our brain is associated with the unconscious and the subconscious. Only a small part is dedicated to the conscious and the awareness, which in turn is also considerably controlled from the subconscious. Learning (e.g., to drive) takes place in the border region between conscious and unconscious.

Technological progress today is often based on the replication and simulation of natural processes and structures. The generic term for this is the bionics. In the media, the medical research of the brain is often reported with the primary goal to cure diseases, for example, in large EU Research programs (such as the Human Brain Project, HBP). There are also other scientific disciplines that strive to analyze the brain in order to drive innovation. Computer professionals try to model the brain in a reverse engineering process and—based on technical solutions to develop practical applications such as voice recognition and other systems—they aim at creating, developing, and perfecting Artificial Intelligence.

Behavioral psychologists and social scientists, on the other hand, have long tried to explore the processes in the brain and the interaction between consciousness and the subconscious with their methods, e.g., to come to a better understanding of the processes of assessment and decision making and the influences on that. From that they are trying to find methods and tools to influence purchasing behavior, group dynamics, leadership decisions, etc. They try to explore which conditions particularly promote creativity and innovation, and thus also often operate in the area of the subconscious mind and its interactions with the consciousness. Viewed in simple analogy, the structures of consciousness and the subconscious are already applied to organizations: mission statements, visions, views of the future of an organization—"Big Pictures," corporate strategies, development of the corporate culture, etc. They aim at anchoring these "Big Pictures" in the subconscious of the people and the parties interested or affected. People are "manipulated" to positively influence the actions of the organization. That's what—mainly—the discipline organizational behavior is all about.

This analogy to the "brain" of the organization, however, may not only be comprehended on the level of the individual person and its capacities and opportunities. Infrastructures, in particular systems and networks, but today also the design of working environments enabling innovation are essential elements of the "subconscious mind" of an organization. These infrastructures—next to mission statements and strategies—too have to be considered as part of the "subconscious mind" of the organization and thereby part of the operation of the brain of the individual person working in this organization. How to model and manage this corporate digital transformation on the one hand and to consider the field of behavioral psychology on the other hand is subject of this book.

The interactions between this "subconscious mind of an organization" and the conscious purposeful actions of employees at all levels of an organization affect the sustainable success of an organization in an increasingly volatile environment. The technological developments in Big Data, Social Media, Artificial Intelligence (AI), Augmented Reality, Internet of Things, and Internet of Everything will only be used successfully by an organization when they are carefully and purposefully woven into the "subconscious mind of the organization" and thus in its infrastructure. Successfully "wired" the new technologies can be made accessible and usable to people acting, and they can support the visions and strategies and the taking of appropriate action in the organization. Thus, the digital transformation is shaping the subconscious mind of our organizations. Finally, we are on the way to collaborate with AI and to "merge" human and AI in some sort of hybrid intelligence, becoming a constitutive part of our future organizations, when innovating our organizations. It is one of the key management tasks to shape our organizations and their subconscious mind actively instead of just "letting it happen" in order to make them fit for the future. So this book addresses managers, consultants, students as well as all people interested in the societal and economic change brought with by the digital transformation that is just happening and that will impact our professional and private life for some time.

Dear reader, this book is to encourage you "to rethink" the organizations where you work, where you have influence, or where you are just interested in. This book is intended to help you to develop ideas and innovations in this respect.

Fictional microstories—though already close to reality—should support you in the understanding of the content and the theoretical descriptions that are partly elaborated—in a sometimes metaphoric style—from analogies to how our brain and our subconscious are working. Together they form three stories of companies in three different business areas (a hospital group, a steel company, and a retailer). Guiding principles for shaping the subconscious mind of organizations provide actionable hints for how to deal with the digital transformation in the respective organization.

First, in Chap. 1—the introduction—the terminology is semantically explained in order to anchor the analogies and metaphors used. In Chap. 2, the relevant foundations of cognitive science and behavioral psychology are presented. Chapter 3 explains the elements of an organization's infrastructure and analyzes the relevant technologies and methods as well as expectations for the future, in particular

Artificial Intelligence and "predictive modeling" along with other topics in the area of Big Data, Smart Data, etc. Chapter 4 develops a model of the "subconscious mind of an organization" and its interfaces to the "conscious mind of an organization." Chapter 5 is devoted to the question of whether, how, and how fast the "subconscious mind of an organization" can be developed and transformed. This chapter provides guidance and guidelines how to design sustainable organizations and what has to be considered with special emphasis on decision making processes. Chapter 6 discusses the digital transformation as a societal meta-development directly impacting organizations. Based upon the subconscious mind of organizations as metaphoric view for the question of structural and process-oriented aspects of organizations as well as for methodological and technological questions of decision support and automation, a next step is suggested: To view an organization or parts of an organization as a hybrid intelligence of human and AI "Hybrid intelligence" serves as a "Big Pictures" of the future—following an obviously inevitable but influenceable development of the cultural evolution of mankind—and discusses some ideas how to approach and manage this in an organization.

Given the rapid technological development—the digital transformation—experiences, observations, and literature, as well as expectations derived from them, provide the foundation for the hypotheses developed and their argumentation and justification. Statistic or scientific derivations and findings are therefore discussed only to a lesser extent.

Notes to the Emergence of the Idea for this Book:
The cybernetic patterns described in Chap. 5 (the amplifying circuit, the balancing circuit, the element of delay, etc.) partly correspond—in an analogy view—to the basic types of control loops, as we know them from the technical discipline of control engineering. In 1992, I applied these cybernetic considerations and principles on manufacturing systems and companies and published it in German in a book with the title (translated) "Manufacturing Execution Systems—principles from control engineering applied on logistics" (Leodolter 1992). At that time, I did not know the book by Peter Senge on "organizational learning" (The fifth discipline, Senge 1990). The thoughts of Peter Senge I discovered in 1995. In an article on adaptive organizations (translated) "Adaptive organizations—a prerequisite for successful and sustainable businesses" (Leodolter 1997), I enriched my ideas from 1992 and thus expanded my reflections on the application of the principles of control engineering on management of entire organizations. In this article, I used the metaphor of the "subconscious mind of organizations" for the first time as well as the comprehensive description of infrastructure—in a wider meaning —as the foundation of organizations. From the analogy that these structures partly form the behavior of organizations, finally the metaphor of the "subconscious mind of organizations" emerged. After years of work in upper and

top management levels (CIO and CEO) in the healthcare industry and the ongoing observation and study of information and communication technologies (ICT) in the course of my professional work as CIO and my teaching position at the Technical University of Graz, and after reading the book of Daniel Kahneman "Thinking fast and slow" and the book "How to create a mind" by Ray Kurzweil, these ideas "popped up" again. The current wave of innovation in the application of ICT in organizations—under the tags Big Data, Visual Analytics, In-Memory Computing, Massive Parallel Processing, Predictive Analysis, etc. summarized as "digital transformation"—now really provides an opportunity to bring these ideas into practice. This forms the background "to rethink" organizations in order to make them more effective, efficient, and sustainable. So in this book, I would like to provide suggestions for improving the effectiveness and sustainability of organizations and complex systems and how to design these—facing the digital transformation. Having the experience of publishing the book in German in 2015, I had many discussions and presentations on this topic. The technologies—especially Artificial Intelligence—went on in their development and application. That is why I reworked the book in English adding another section with more emphasis concerning the digital transformation and discussing the collaboration and merging of humans and AI in organizations as "hybrid intelligences."

You have 3 ways to read this book:

- The recommended option is, of course: read from front to back—that is the best way to recognize and comprehend the extensive interrelationships—backed by the microstories as examples. In Chap. 2—considerations based on behavioral psychology and cognitive science—and in Chap. 3—the elements of an organization's infrastructure and the relevant existing and expectable technologies—concepts, theories, and technologies are introduced in order to lay a foundation for a thorough understanding of "the subconscious mind of organizations." So readers who are already familiar with cognitive processes and psychology or technologies or both can go through the respective chapter more quickly. But they should not skip them completely because they also provide numerous "links" to the concept of the "subconscious mind of organizations," which is briefly sketched in the introduction and elaborated as model in Chap. 4.
- To get a taste (Quick Scan)—or to remember after complete reading—the ideas are also conceivable from the fictitious microstories about a hospital group, an industrial company, and a commercial enterprise as presented in the book. Put together or read sequentially, they each provide a consistent story from the respective industry.

- The third option is fast reading—in particular focusing on the design of the "subconscious mind of organizations" by reading the summaries of the chapters and the "arrowed" guiding principles in the text. This is a more abstract access to the matter. This is also well suited to call back to mind what you read—possibly after some time.

For those readers who choose to start with the stories and/or the guiding principles, I hope that their interest is aroused so much that they will read the whole book and even remember what this book was about in 5 years, because they use this conceptual idea of the "subconscious mind of organizations" as framework or fundamental hypothesis when thinking about organizations and decision making. As you all know about your personal subconscious mind, it is easy to memorize a metaphor and remember it when needed.[1]

Graz, Austria   Werner Leodolter

---

[1]For the references to these notes, see Chap. 1.

Acknowledgements

Thank you, of course, first to you, dear readers, you bought this book and maybe already have read it. Thanks to those who have made the book, especially to Stefan Einarson from Springer for your suggestions and corrections.

Thanks to Reinhard Krepler, who, as a manager and scientist, has motivated me to summarize my publications and my contributions to science and teaching. Thus, he initiated a thought process with me, in the course of which the idea to this book came.

Thanks to all the colleagues who accompany me in my professional life, who have borne my temporary impatience and provided and initiated ideas, without knowing it. Ideas, which have shaped my professional activities. Numerous situations and episodes with these colleagues contributed in the context of this theme in the course of writing. These 35 years intense and varied professional experience and my passion to read a lot have built the foundation, from which the idea and conception as well as the contents of this book emerged.

Many thanks to Alfred Gutschelhofer, who, as a university professor and scientist, read the manuscript in a first German version. He has initiated further ideas and has encouraged me to publish the German book 2015. Thanks to Richard Ackbar who did the proofreading as a native speaker.

Thanks to my family, especially to my dear daughters Martina and Nora and their partners Klaus and Mario for their reading of the German book and their valuable suggestions. My deepest gratitude for my dear wife, Andrea, who had a lot of patience with my frequent absence and provided the love and familiar backup and support for my career and professional work in the last decades as well as for my work as author. Without Andrea, this book would not exist.

Contents

About the Author

Univ. Prof. Dipl.-Ing. Dr. techn. Werner Leodolter is CIO of KAGes. Previously, he was 26 years in management positions in information management in various industrial companies and in the Styrian hospital Company (KAGes). From 2008 to 2013, he led the KAGes (an Austrian hospital group with 17,000 employees) as CEO through a period of organizational realignment. He is also a member of Telehealth Services Commission of the Federal Ministry of Health. He is a University Professor of Applied Management Studies in Health Care at the University of Graz (KFU) and lecturer on the subject of Information Management at the Technical University Graz (TUG). He is also author of the book "The subconscious mind of organizations: New technologies—Thinking organizations new," published in summer 2015 in German.

Chapter 1
Introduction

Abstract The concept of the subconscious—derived from the behavioral psychology and cognitive sciences—and the skills for intuition and for intuitive action are closely linked. The analogue observation in the sense of the "subconscious mind of organizations" is appropriate and is practically indicated with the arrival of the new technologies in the organizations. Metaphorically the infrastructures, as well as structure and processes of an organization—together with the values, attitudes and strategies of the organization—shape the subconscious mind of an organization which in turn influences decision making as the central skill of each organization. This subconscious mind is massively affected by what we call the digital transformation and hybrid intelligencies that emerge from the coexistence, new collaboration and the merging of man and machine—hopefully with humans staying in the driver's seat.

Keywords Organization · Behavioral psychology · Cognitive science · Subconscious · Infrastructure · Decision Making

First, the concepts of the subconscious and its differentiation from awareness and perception are set. Let's start with Sigmund Freud, the father of Psychoanalysis (following Sandler et al. 2003). Sigmund Freud presented two different and not entirely congruent models of the psychological apparatus:

- First, the model conscious/preconscious/unconscious (topographic model)
- later, the model Me/It/superego (structural model of the psyche).

Let us consider the topographical model: conscious, preconscious and unconscious: The unconscious is the region of the human psyche, which in psychology is not directly accessible for the consciousness. In normal usage it is often called "subconscious". Depth psychology assumes that unconscious mental processes influence human action, thought and feeling significantly, and that becoming aware of unconscious processes is essential for the treatment of neuroses. In contrast, all mental processes and content that is not activated at the moment—but contrary to the preconscious are in principle accessible in case of need—are summarized and designated as the subconscious mental region, that can be re-activated any time (see

© Springer International Publishing AG 2017
W. Leodolter, *Digital Transformation Shaping the Subconscious Minds of Organizations*, DOI 10.1007/978-3-319-53618-7_1

Roudinesco and Plon 2004). In this book the subconscious and the unconscious are summarized in the term "subconscious mind".

The concept of intuition is relevant in this context too:

Intuition is "the ability to acquire insights into issues, perspectives, laws or the subjective consistency of decisions without discursive use of reason" (Gigerenzer 2007). The ability to detect complex characteristics and motions unconsciously, consciously and instinctively in a split second is—from an evolutionary point of view—an ability and an attitude that enables us to make a distinction between friend and enemy on the one hand and to decide for a fight or flight reaction in a split second on the other hand.

This book is about organizations and hypothesizes, that it is allowed to speak about the subconscious mind of organizations—at least metaphorically or in form of an analogy.

Referring to Coenen (2002) analogies are described as follows: The analogy as rhetorical term refers to a stylistic method in which similar structures or situations are placed in a context. There is an analogy between two things, when they are similar in a feature, even if they differ in other features. This stylistic method is often employed, when an already known information or an already found consensus from a comparable factual connection, or a comparable context are used for illustrating or for amplifying an argument in a different context. If we draw tangible conclusions for the new, comparable factual connection from the context otherwise already known, we may speak of an analogy-based conclusion.

Hofstadter et al. explain in their book "The analogy—the heart of thinking" (2014) some essential features of the analogy as an essential feature of intelligence: The concepts in which we are thinking, are due to the language. Errors allow us a look into the subconscious because in the mind words always compete with each other. The new, that we perceive, we apply to what we already know. Analogies are pivotal to all our thinking. Intelligence is the ability to very quickly recognize deep analogies and the ability to put the finger on the essentials of a situation within a short time. Mathematicians also let themselves be guided by vague intuitions and insights—only afterwards they justify their actions then and fill in the gaps (e.g. with logic). Analogy and intuition are quite similar.

The area to which we want to apply the analogy, is the world of organizations of all kinds. Let us consider therefore the concept of organizations.

The Gabler Dictionary of Economics designates an organization as the formal rules of the division of labor in a system, but also draws attention to informal arrangements, which also can be effective. Generally, the concept of organization is not clearly defined. The definition depends on the respective underlying organizational theory approach (see Springer Gabler Verlag (Hrsg) 2010).

The aforementioned informal arrangements are part of the "Subconscious Mind of Organizations" and are mutually interdependent with the "conscious" formal rules. In this respect, the "subconscious mind" in its everyday operation—although unconsciously for the organization—can be shaped to a large extent.

Let us therefore consider the term organization in this book very openly and widely as a "purposeful connection of individuals and organizations with a form of co-ordination and/or self-regulation to accomplish that purpose of the organiza-tion". The term "organizations" therefore includes on the one hand small and medium sized enterprises (SMEs) up to large corporations in the profit sector but also in the non-profit sector and—secondly—in communities and public entities on the other hand. This book focuses primarily on organizations in the profit and non-profit sector, although some ideas and concepts may also apply on commu-nities or public entities as organizations. The given scenarios cover SME's (like the local retail company) as well as larger entities (like the steel company) up to a hospital group as big organization.

For the theme of this book an acceptable and workable definition and structure of the human layers of consciousness as a basis for analogies towards organizations can be found in the work of psychotherapist Staroversky (2013). He structures the psyche in 3 layers, a structure that will be the foundation for the following considerations:

- the conscious mind—Consciousness
- the subconscious mind or the preconscious mind—the subconscious
- the unconscious mind—the unconscious, which is considered in this book as "sub-domain" of the subconscious mind.

Another term which defines and represents a particular form of analogy and is of relevance for this book, is the bionics. From an abstract perspective the bionics is a form of analogy from natural phenomena to technological constructs and concepts. Bionics (also biomimicry, biomimetics or biomimese) deal with the transferring of phenomena from nature to technology. The oldest known example is Leonardo da Vinci's idea of transferring the flight of birds to the design of flying machines. The most common example of the modern life is the velcro fastener inspired by bur-docks. The Bionics is based on the assumption that nature develops and optimizes structures and processes through evolutionary processes of which mankind can learn from (see Blüchel 2006). As an interdisciplinary research field, bionics attracts scientists, engineers, architects, philosophers and designers. The bionics is about systematically detecting solutions of living nature; it distinguishes itself from the purposeless inspiration by nature. The goal of bionics is always a technical object or method separate from the natural. Thus the bionics differ from sciences that use and enhance biological processes, such as bioinformatics, biophysics and biochemistry.

In this respect the analogy from the human brain and its levels of consciousness towards organizations is quite appropriate.

An important question is: Are there already applications of the concept "sub-conscious mind" on organizations in literature?

An early approach was—if not based on an organization but viewing beyond the individual—the assumption of the "collective unconscious" by Carl Gustav Jung:

The "analytical psychology", founded by Carl Gustav Jung, has many similar-ities to Freudian psychoanalysis. Both describe the possibility of (re)uncovering of the unconscious through psychodynamic therapy. An innovative element of the theory is the assumption of a "collective unconscious" by Jung. Unlike Freud, Jung saw the collective unconscious parallel as a kind of reservoir of experiences which mankind had internalized in the evolutionary process. The archetypes described by Jung in 1919 become manifest in the collective unconscious. The most important are: Animus and Anima (the image of the male and the female), shadow (negative, socially undesirable, suppressed personality traits) and the old wise male and female. Then there are the symbols of the self as a comprehensive expression of wholeness of the psyche. Archetypes are the pre-existent unconscious form of the individual archetypes, which determine the individual psyche. A human being emerges from the collective unconscious and its subjectivity is produced through a process of progressive integration of animus and anima, Jung designated this as individuation (see Roudinesco and Plon 2004). There are numerous seminars and literature on the use of psychoanalysis in organizations—but primarily on the influence of the organization on the individual. They focus on the ways of influencing and understanding individuals in their roles in the organization—they deal with the application and use of psychoanalysis on the people in the organization and on the customers of the organization for the purpose of the respective organization. Lohmer (2014), for example, deals with the psychoanalysis of organizations and propagates the "psychodynamic organizational consulting". The discipline of organizational behavior mainly concentrates on how individuals are influenced in their behavior by the organization in order to deliver a good performance for the organization.

I have used the concept of "the Subconscious Mind of Organizations" in a scientific article (Leodolter 1997) in connection with the establishment of learning organizations and the necessary infrastructures for that.

The massive developments in the information and communication technologies (ICT) in the past two decades have induced a fundamental change: contrary to the early stages of ICT use, when ICT tools were mainly experienced in the organizations with the tools for the respective jobs, now individuals (as parts and employees of the organizations) bring in their experience with advanced tools themselves. Thus they transfer their personal contact with new technology and no longer are only users of the tools provided by the organization as in times before. They urge for a "user experience" as Apple, Facebook, Google, the game industry etc. usually offer. Organizations also increasingly offer such user experience—at least they try to increase their efforts in order to attract talents. The "war for talent" is already in progress as the press and literature tell us. The demographic trends over the next decade in the western industrialized countries will enforce the expected "war for talent".

A word often used in this context is "infrastructure"—a term for the totality of the information, communication and organizational resources as well as its organizational design and embedding.

The term infrastructure is primarily used for public buildings and facilities, supply facilities, transport etc. but also for basic technical facilities in the private sector, for example in companies. So infrastructure management may include management of roads, buildings and basic technical services, such as power or communication in industrial parks or business infrastructure. Specific forms of organizations of the private sector for the management of such investments are often referred to as "(on-) Site Management" or "Facility Management". In the companies themselves the term "IT infrastructure" has been enforced in recent years.

The IT infrastructure is the set of all IT-related buildings, communication services (network), equipment (hardware) and programs (software), provided for automated information processing on a higher level by a subordinate level (lat. infra "Under"). The higher level does not have the possibility of directly influencing the processes of the lower level; the planning authority for these processes remains at the lower level.

The term "consolidation of the IT infrastructure" refers to the process of unification and merger (or even fusion) of server or desktop systems, applications and databases or strategies (see TecChannel Blüchel 2006). The goal is usually to achieve a simpler and more flexible IT infrastructure—often by reducing physical systems and replacing those by virtual systems.

The subconscious mind of organizations within the meaning of the statements in this book is therefore—in the form of an analogy to the subconscious and the consciousness of the individual and the concepts of psychoanalysis—a socio-technical construct consisting of

- technical infrastructure (which allow and support subconscious and conscious action for the organization) and
- structures and processes of an organization (formal and informal rules) and
- values, attitudes and strategies.

That is—so to say—the "constitutional" view of the subconscious mind of an organization.

The subconscious mind of organizations provides the framework for conscious and unconscious actions of individual employees and executives and significantly form and influence them. The performance and development of an organization arises from the countless decisions taken either by individual employees and executives more or less consciously. They are possibly supported by systems based on parameters and algorithms or even decisions taken by the system itself.

So the—so to say—"behavioral" view of the subconscious mind of organizations can be described as what is perceived from the respective level and position in the organization from the other parts of the organization and its environment, in particular the response to control measures from the respective management level.

The analogy from the human brain and its levels of consciousness towards organizations is therefore quite appropriate—also from evolutionary aspects. Let us have a look at two concepts related to this idea:

- Bionics is based on the assumption that evolutionary processes optimize structures and processes—features mankind is learning from.
- Carl Jung saw the collective unconscious as a kind of reservoir of experiences which humanity have internalized parallel to its evolution.

The research priorities in the EU (HBP, the Human Brain Project) and the US (Brain Initiative) and the entire research and development field of Artificial Intelligence are also an expression of this evolutionary development.

In the next chapter, the cognitive science and behavioral psychology as a basis for decision-making are brought into the spotlight. This is followed by a chapter that elaborates the infrastructural and technical aspects which characterize the subconscious mind of organizations and the interface to their consciousness. These aspects of technology have the potential to form and shape the subconscious mind of organizations. Dear reader if you are already familiar with cognitive processes and psychology you can go through the following chapter more quickly. But please mind that the following chapter will also provide "links" and first short interpretations concerning the above briefly sketched concept of the subconscious mind of organizations.

Literature

Blüchel KG (2006) Bionik. Wie wir die geheimen Baupläne der Natur nutzen können. Goldmann, München. ISBN 3-442-15409-X

Coenen H-G (2002) Analogie und Metapher. Grundlegung einer Theorie der bildlichen Rede. Berlin

Gigerenzer G (2007) Bauchentscheidungen. Die Intelligenz des Unbewussten und die Macht der Intuition. Bertelsmann, München. ISBN 978-3-570-00937-6

Hofstadter D, Sander E, Held S (2014) Die Analogie: Das Herz des Denkens, 1. Aufl. Tropen-Verlag (Verlag C.H. Beck im Internet). ISBN 978-3-608-94619-2

Leodolter W (1992) Fertigungsleitsysteme – organisatorische und informationstechnische Leitlinien. B.G.Teubner, Stuttgart

Leodolter W (1997) Lernfähige Organisationen – eine Voraussetzung für nachhaltig erfolgreiche Unternehmen. Der Wirtschaftsing, 37(1), 27–31

Lohmer M (2014) Psychoanalyse in Organisationen: Einführung in die psychodynamische Organisationsberatung. Kohlhammer, Stuttgart.

Roudinesco E, Plon M (2004) Jung, Carl Gustav. In: Dictionnaire de la Psychanalyse, 1997. Aus dem Französischen von Christoph Eissing-Christophersen u. a. Wörterbuch der Psychoanalyse. Springer, Wien, pp S 510–515

Sandler J, Holder A, Dare C, Dreher AU (2003). Freuds Modelle der Seele. Eine Einführung. Psychosozial, Gießen

Senge P (1990) The fifth discipline, the art and practice of the learning organisation. Doubleday currency

Springer Gabler Verlag (Hrsg) (2010) Gabler Wirtschaftslexikon, Stichwort: Organisation. http://wirtschaftslexikon.gabler.de/Archiv/773/organisation-v6.html. Zugegriffen: 13. Juli 2014

Staroversky I (in www.staroversky.com; Three Minds: Consciousness, Subconscious, and Unconscious, Posted on May 23rd, 2013) mit Bezugnahme auf Corsini RJ, Wedding D (2011). Current psychotherapies, 9. Aufl. Brooks, Belmont

TecChannel.de Leitfaden zur Server-Konsolidierung. 17. Jan. 2006. Accessed 5 Aug 2010

Chapter 2
Considerations Based on Behavioral Psychology and Cognitive Science

Abstract Subconscious processes, which are based on experience and which can be trained, control our behavior and follow an "internal reasoning". They are shaped—among other things—by linguistic constructs. These subconscious processes form the "fast and intuitive thinking". The benefits of this efficiency are associated with hazards such as prejudices, bias etc. Intuition is an essential resource of our cognitive system and can be influenced. The subconscious mind acts as an associative machine which works on the basis of pattern recognition. So we form our reality in a three-step process consisting of suppression, distortion and generalization. Associations and analogies are essential characteristics of intelligence. These aspects from behavioral psychology and cognitive science at the personal level can be used to support good decision-making and—in the context of the respective organization—they also can be used in the light of the metaphor of shaping "the subconscious mind of organizations" in order to cope with change, uncertainties and disruption.

Keywords Intelligence · Intuition · Motivation · Perception · Cognition · Pattern recognition · Decision making

When we want to contemplate the subconscious mind of organizations it is important and permitted to consider the subconscious mind of the involved actors of an organization first—of managers, of employees, of stakeholders, of the humans in their relevance for decision making mechanisms. Here we meet cognitive sciences and behavioral psychology. As we will see later, the extensive consideration of these issues is valuable while applying this on organizations and their subconscious mind and conscious mind in an analogy.

© Springer International Publishing AG 2017
W. Leodolter, *Digital Transformation Shaping the Subconscious Minds of Organizations*, DOI 10.1007/978-3-319-53618-7_2

2.1 "Mind" and "Self"

The psychologists and neuroscientists, such as Antonio Damasio in his book "Self Comes to Mind" (2012) talk about the awareness and the large area of the unconscious. According to the above definitions, it is permissible—even in terms of behavioral psychology in the social and economic context—to define the realm of the unconscious, which is close to consciousness, as the subconscious mind. In the behavioral and psychological considerations in social and economic context it's all about the particular interaction and mutual influence of the conscious and the subconscious.

In the mainly English literature from this area we also repeatedly encounter two concepts which I initially want to explain: "mind" and "self".

"Mind"—sometimes synonymously described as "spirit" or "soul"—can be seen as a collection of cognitive abilities, skills, processes and properties that influence and allow consciousness, perception, understanding, thinking, judging and memory —applicable to people (the human mind, the human soul) and other forms of life. The "conscious mind" or "consciousness"—awareness—and the "subconscious mind"—the subconscious—are instances of this term. Psychology considers and treats all of the conscious and unconscious processes and actions. The subconscious also includes elements of the unconscious, such as values, beliefs, patterns, and a specific personal (subjective) map of reality—a model of reality.

"Self"—the self is a diversely used term with respect to psychological, socio-logical, philosophical and theological meanings—the sensation of being a unified, consistent, sentient, thinking and acting being in introspective sense. The self serves the reflection, reinforcing and emphasizing of the concept of "Me" (see Bracken 1996). C.G. Jung refers to the symbols of the self as a comprehensive expression of the totality of the psyche.

The cognitive scientist Damasio (2012) described the development of consciousness from evolution: Consider the awareness of animals as the processing of the images of the environment, which is carried out based on certain internal images. The self thus focuses the process of consciousness in dangerous situations. The dualism body-mind in such situations is also discussed in philosophy—philosophy discusses that specifically as "mind-body problem".

"Mind" and "self" can—in analogy—also be applied to organizations concerning the formal as well as informal rules and the corporate culture on the one hand and the reflective, reinforcing and emphasizing processes focusing on the organization's mission and business on the other hand—with the potential to foster entrepreneurship. The concepts of organizational behavior form a prominent research area in the US but are less prominent in the german speaking countries and mainly deal with the way employees are and can be influenced. It also deals with the influence and interrelationships of leadership, strategy, vision and mission etc. challenging how the mind, the behaviour and the performance of humans in an organization can be influenced.

2.2 Processes of Consciousness

When—in the course of evolution—the processes of consciousness have become more complex due to the benefits in evolutionary competition, functions such as memory, reasoning and language developed. It was possible to survey the near future, to delay reactions and responses—automatically proposed by the subconscious—in order to prevent or to enable delayed reaction and to allow not only an immediate reward and punishment, but the estimate of future benefits and drawbacks. This was also the beginning of the development of a socio-cultural homeostasis ("equilibrium") and therefore of civilization, culture, art, science and research up to the contemporary social and technological achievements (see Damasio 2012). Applied on organizations it is interesting to evaluate the controlling processes on the respective level of an organization from this point of view: What has to perform in the realm of the conscious mind of an organization and what can be left to the subconscious mind?

2.3 Learning and How to Build a Mind

As learning is a key feature in Artificial Intelligence (AI) and as learning has an eminent role in forming one's subconscious mind—and presumably also the subconscious mind of organizations—, it seems appropriate to look at this issue more closely from the viewpoint of cognitive sciences:

How do intelligent minds learn? Consider a child navigating through its day, bombarded by a kaleidoscope of experiences. How does its mind discover what's normal and begin building a model of the world? How does it recognize unusual events and incorporate them into its worldview? How does it understand new concepts, often from just a single example? How does its mind emerge?

These are the same questions machine learning scientists ask as they are moving closer to AI that matches or even beats human performance. Much of AI's recent victories are rooted in network architectures inspired by multi-layered processing in the human brain (Fan 2016).

Kumaran et al.—as scientists from Google DeepMind and Stanford University—published an update on what is the actual theory of how humans and other intelligent animals learn. Following this review paper in *Trends in Cognitive Sciences* (Kumaran et al. 2016) the Complementary Learning Systems (CLS) theory (McClelland et al. 1995) states that the brain relies on two systems that allow it to rapidly soak in new information, while maintaining a structured model of the world that's resilient to noise. Applied to organizations this concerns the ever challenging question of what is really relevant for my organizations and what can be neglected as "noise".

Given the discrepancy found when studying patients with damage to their hippocampus, it can be reasoned that new learning and old knowledge likely rely on two separate learning systems with the hippocampus being the site of new learning, and the cortex—the outermost layer of the brain—as the seat of remote memories. According to CLS, the cortex is the memory warehouse of the brain. Rather than storing single experiences or fragmented knowledge, it serves as a well-organized scaffold that gradually accumulates general concepts about the world—it is learning.

Experiments with multi-layer neural nets, the precursors to today's powerful deep neural networks, proved that: Experiments with artificial learning systems showed that they gradually learned to extract structure from the training data by adjusting connection weights—the computer equivalent to neural connections in the brain. Thus the layered structure of the networks allows them to gradually distill individual experiences (or examples) into high-level concepts.

Similar to deep neural nets, the cortex is made up of multiple layers of neurons interconnected with each other, with several input and output layers. It readily receives data from other brain regions through input layers and distills them into databases ("prior knowledge") to draw upon when needed. "According to the theory, such networks underlie acquired cognitive abilities of all types in domains as diverse as perception, language, semantic knowledge representation and skilled action," the authors wrote.

Perhaps unsurprisingly, the cortex is often touted as the basis of human intelligence. Yet this system isn't without fault. For one, it's painfully slow. Since a single experience is considered a single "sample" in statistics, the cortex aggregates experience over years in order to build an accurate model of the world. Information stored in the cortex is relatively faithful and stable. It's a blessing and a curse—we need time to learn. Related to Artificial Intelligence networks that means that jamming new knowledge into a multi-layer network, without regard for existing connections, results in intolerable changes to the network. The consequences are so dire that scientists call the phenomenon is "catastrophic interference" (Fan 2016).

Thankfully, we have a second learning system that complements the cortex. Unlike the slow-learning cortex, the hippocampus concerns itself with breaking news. Not only does it encode a specific event, it also registers the context in which the event occurred. This lets us easily distinguish between similar events that happened at different times.

Following Kumaran et al. the reason that the hippocampus can encode and delineate detailed memories—even when they're remarkably similar—is due to its peculiar connection pattern. When information flows into the structure, it activates a different neural activity pattern for each experience in the downstream pathway. Different network pattern; different memory.

In a way, the hippocampus learning system is the antithesis of its cortical counterpart: it's fast, very specific and tailored to each individual experience. Yet the two are inextricably linked: new experiences, temporarily stored in the hippocampus, are gradually integrated into the cortical knowledge scaffold so that new learning becomes part of the "database".

But how does that work? Scientists don't yet have all the answers, but the process seems to happen during rest, including sleep, the type of electrical activity that propagates to the cortex is called short wave ripples (SWR) (McClelland et al.

1995). When examined closely, the ripples were actually "replays" of the same neural pattern that the animal had generated during learning, but sped up to a factor of about 20. Picture fast-forwarding through a recording—that's essentially what the hippocampus does during downtime. This speeding up process compresses peaks of neural activity into tighter time windows, which in turn boosts plasticity between the hippocampus and the cortex.

Following Kumaran et al. in this way, changes in the hippocampal network can correspondingly tweak neural connections in the cortex. Unlike catastrophic interference, SWR represent a much gentler way to integrate new information into the cortical database. Replay also has some other perks. You may remember that the cortex requires a lot of training data to build its concepts. Since a single event is often replayed many times during a sleep episode, SWRs offer a deluge of training data to the cortex.

SWR also offers a way for the brain to "hack reality" in a way that benefits the person. The hippocampus doesn't faithfully replay all recent activation patterns. Instead, it picks rewarding events and selectively replays them to the cortex. This means that rare but meaningful events might be given privileged status, allowing them to preferentially reshape cortical learning.

"These ideas and this view on memory systems are being optimized to the goals of an organism rather than simply mirroring the structure of the environment," the authors explained in the paper (Kumaran et al. 2016). This reweighting process is particularly important in enriching the memories of biological agents, something important to consider for Artificial Intelligence too.

The two-system set-up is nature's solution to efficient learning. "By initially storing information about the new experience in the hippocampus, we make it available for immediate use and we also keep it around so that it can be replayed back to the cortex, interleaving it with ongoing experience and stored information from other relevant experiences," says Stanford psychologist Dr. James McClelland. CLS has been instrumental in recent breakthroughs in machine learning (Kumaran et al. 2016).

I am convinced that these ideas and concepts might be considered when thinking about and designing and shaping the "subconscious mind" of organizations as well e.g. by emulating a "two-system set-up" by providing enough time and resources to make simulations of alternative decisions to be taken or to implement e-learning and micro-learning—possibly blended with simulations.

But let us first get an idea of how Artificial Intelligence is using these concepts "written by nature" [according to Fan (2016)]: Convolutional neural networks (CNN), for example, are a type of deep network modeled after the slow-learning neocortical system. Similar to its biological muse, CNNs also gradually learn through repeated, interleaved exposure to a large amount of training data. The system has been particularly successful in achieving state-of-the-art performance in challenging object-recognition tasks. Other aspects of CLS theory, such as hippocampal replay, has also been successfully implemented in systems such as DeepMind's Deep Q-Network.

"As in the theory, these neural networks exploit a memory buffer akin to the hippocampus that stores recent episodes of game play and replays them in inter-leaved fashion. This greatly amplifies the use of actual game play experience and avoids the tendency for a particular local run of experience to dominate learning in the system," explains Kumaran. "We believe that the updated CLS theory will likely continue to provide a framework for future research, for both neuroscience and the quest for artificial general intelligence", he says.

To use this knowledge about learning, memory and experience for metaphoric views and source of analogy for considerations in management, leadership and everyday decision making seems highly relevant for individuals as well as for organizations as socio-technical systems.

So let us follow these ideas of 2 systems collaborating, of blending of memory and experience, of fast and slow and of building concepts as part of our minds later when we discuss what has to be considered when forming and shaping the sub-conscious mind of an organization (Chap. 5) and in particular when thinking about hybrid intelligencies (Chap. 6).

2.4 Decision-Making and Learning Processes

With progressive evolution of organisms, the subconscious decision-making pro-cesses could be continually managed and controlled in a better way and in particular the decision-making processes of humans improved through learning. There are two types of control: the conscious control and the unconscious control, which is partly developed under conscious control. In human childhood and adolescence there is a lot of time available to condition the subconscious processes in accordance with the conscious objectives and cultural conventions. Parts of conscious control are—figuratively speaking—outsourced to a "Subconscious Server", and the control is taken over by the consciousness in "special" situations as Suhler and Churchland (2009) have convincingly shown.

Subconscious processes are thus becoming a suitable ability to control behavior and allow the consciousness more time and more resources for further and chal-lenging tasks such as analysis, deliberation and decision-making. The consideration and decision-making however often are subject to numerous "biases", "priming" and "framing"—biologically determined or culturally acquired prejudices, imprints and frames. The conscious control of awareness can reduce these negative effects, but requires attention and thus cognitive resources. Ap Dijksterhuis, a Dutch psychol-ogist has demonstrated in an experiment with purchasing decisions for normal consumer goods on the one hand and major buying decisions (house, car) on the other hand that also unconscious processes are subject to "internal reasoning" and can lead to good results if the subconscious mind is well prepared by past experi-ences and training. There are not always exact considerations of possible advantages and disadvantages necessary (see Dijsterhuis 2006). This experiment demonstrates how powerful the right combination of subconscious and consciousness can be. The decision is made consciously and efficiently at a reasonable cost, which takes place in the subconscious as some sort of "due diligence".

Applied on organizations this emphasizes the importance of self-control and self-motivation of the people responsible of the teams and the organizational units. Thus to set the right goals, to establish good processes, an agile organizational structure and the appropriate incentives are key challenges to shape the subconscious mind for your organization.

If we practice something enough, we store skills, knowledge and expertise in the subconscious and are mostly unaware of the technical execution steps. I remind you of driving a car, as described in the preface, or the elite among musicians who—consciously focus on the expression and thus setting themselves apart from their competitors—with a good part of technical perfection "outsourced" to the subconscious. In a metaphoric view these considerations are also highly relevant for organizations as socio-technical systems.

2.5 The "Extended Mind Thesis"

Let us now "extend" the view: The "extended mind thesis" (EMT) suggests that some objects in the environment of a person are used by its consciousness in a way, that these objects can be considered as an extension of consciousness itself. The thesis goes back to Andy Clark and David Chalmers from (1998).

They have argued, that the assumption that our consciousness is "limited by cranial bones" is an arbitrary one. The separation between mind, body and environment is questioned. Because external objects and sources play an important role in cognitive processes, mind and environment, they are seen as coupled systems. It is seen as a key differentiator to traditional processes of interaction with the environment, when external objects are used in cognitive processes as part of an extended cognitive system. Thereby they pursue the same objective as the internal cognitive process. The critics argued, among other things, that this approach would lead to an exuberant, "inflated" concept of knowledge, because for example, a calculator would be seen as such an extension of the cognitive system.

As a result of scientific debate, a moderate revision of the "extended mind theory" developed, which relativizes the assertion of equivalence and complementarity of internal and external elements of cognitive systems and operations. The "extended mind thesis" has thus been granted an explanatory character and value in the consideration of cognitive processes instead of being regarded as part of the nature of mind and knowledge.

This discussion refers to the skills and cognitive processes of people. But it also shows that—in extension and in analogy to this modified concept of the "extended mind thesis"—the metaphoric approach of this book applying the analogy between the consciousness and the subconscious of human and the "conscious and subconscious mind of organizations" is a valid and consistent approach.

Organizations designed and developed by humans can be seen as part of the evolutionary process of mankind. The cognitive processes as there are perception, recognition, decision etc. are being changed by new technologies like augmented reality, visual analytics, decision support etc. The "cognitive" processes of organizations and organizational units as group of individuals collaborating in an organizational setting are even more undergoing massive change with the availability and the immersion of those new technologies. That can be considered as part of the "evolutionary process" of our organizations we work and live in and of society as a whole. We also see this tremendous extension of our organizations' "conscious and subconscious mind" in the increasing connectedness concerning cross enterprise business processes enabled by these ever more sophisticated and highly automated tools. More and more is bound to be part of the subconscious mind, which consequentially has to be shaped very carefully in order to prevent unintended effects.

The value of this approach should be reflected on the one hand in the design of this subconscious mind of organizations and on the other hand in the approach of "Rethinking organizations". This will be detailed in Chap. 6 referring to so called "hybrid intelligence" as a concept for close collaboration between humans and machines.

2.6   Language and (Artificial) Intelligence

Language is one of the essential means of expression of our mental processes. On the other hand language and linguistic constructs characterize our thinking and are therefore part of our consciousness, but also of our subconscious.

Language and its representation in computer science also significantly shapes the development of information and communication technologies. Noam Chomsky was in this respect one of the pioneers of linguistics. Minsky (1988)—by the Society of Minds Theory—very early built the bridge towards the concepts of Artificial Intelligence. See the following two citations from Marvin Minsky in "The society of mind": "Minds are simply what brains do" and "When intelligent machines are constructed, we should not be surprised to find them as confused and as stubborn as men in their convictions about mind-matter, free will, and the like" (Minsky 1988).

Alan Turing very early (about 1950) defined the Turing test as a benchmark for the assessment of whether or not Artificial Intelligence had been achieved. That too is an issue close to the Linguistics: A computer and a human interact under observation by a third party in the role of observer. The observer has to determine by questions, who is the man and who is the computer. If the observer is not able to reliably determine who the computer is, the Turing test (see Turing 1950) is considered satisfactorily passed. In science and near popular scientific use it is permissible to consider systems such as Apple's Siri or IBM's Watson and similar advanced developments to be "Artificial Intelligence".

Aside from this technical importance of language as part of actual and future Artificial Intelligence, language and terminology are a decisive part in professional collaboration in and across organizations as basis for mutual understanding—thus also forming and shaping the subconscious mind of organizations. So the question in this book is how to shape this and how man and machine (AI) can collaborate in "hybrid intelligencies" with the human staying in the "driver's seat".

2.7   "System 1" and "System 2"—Thinking Fast and Slow

Kahneman (2011), the Nobel Laureate in Economics has outlined the basic principles of decision-making with a focus on the economic sector in his book "Thinking Fast and Slow". He introduces the subconscious and consciousness in terms of the behavioral psychology to the decision-making processes in the economic sector by structuring System 1 (Thinking fast) and System 2 (Thinking slow).

System 1 as the fast system represents much of what we summarize under the term intuition. It is inter alia characterized by the following characteristics:

- Operates automatically and quickly
- Requires little or no effort
- Has no conscious, self-chosen management and control
- Develops surprisingly complex patterns of ideas
- Its instinctive skills we have in common with some of the wildlife
- Activities are exercised faster and run automatically
- Knowledge is in memory and is used without deliberate influence and without difficulty (the "preconscious")
- Is subject to biases and influences
- Cannot be "turned off"
- Is subject to cognitive illusions.

System 2 as the slow, consciously working and thereby effort- and attention-requiring system is characterized by the following characteristics:

- Normally operates in a comfortable "standby mode"
- System 1 continuously generates suggestions to System 2, this forms then opinions etc.
- Is activated when an event is detected, the mental models are violated, which in turn are maintained and developed by System 1
- Assigns attention to difficult and laborious mental activities
- Is often perceived as subjective experience of reasoning, choice and focus
- Formulates thoughts
- Has the ability to change the way System 1 works
- The capacity of attention is limited
- Is responsible for self-control—but with limited capacity

- Has the final say in the decision
- It is easier to see the mistakes of others than your own
- Everything that consumes capacity of this "memory" reduces our ability to think.

So the benefits and risks of intuition as an essential element of the decision making become clear. Applying this concept to organizations in an analogy leads us to a better understanding of the way organizations work and to numerous ideas how to shape the subconscious mind of organizations facing the new emerging technologies especially in the border region of the conscious und the subconscious of your organization.

2.8 Decision Making—Cognitive Typologies

The self as a parent concept of psychological description of an individual is structured by Kahneman (2011) in two types: the "remembering self" and the "experiencing self". The experiencing self "leads the life" and would, for example, try to prevent pain or at least keep the phase of pain sensation as short as possible. The remembering self would choose which were the positive and less negative connotations in the memory. In remembrance of the duration of a painful phase for the intensive though short moment of achievement for example, the painful phase gets easily "forgotten". People identify more with the remembering self and try to establish "their story".

So Kahneman derives two typologies in the decision-making process: the fictitious "Econ" as living in the theory of rational decision, as the Chicago School of Economics with Milton Friedman and the laissez-faire approach under the postulate of freedom of choice ("free to choose") has characterized it. On the other hand, he characterizes the human, living in the real world as the "Human", which can not be rational, which is repeatedly exposed to the traps of intuition. The "Human" can be greatly influenced by the presentation of decision documents and products. In these different typologies the dilemma of behavioral economics becomes clear, namely that the "Humans"—especially customers and consumers—need protection against the exploitation of their weaknesses (especially System 1). The behavioral economics does not believe the total rationality and puts questions whether to establish protection mechanisms, such as "Unsafe text" in small print or opt-out rules rather than opt-in rules (for example, Electronic Health Record in Austria, Obama Care, etc.).

Apart from these general considerations and systematization and structuring on the path to good decisions numerous dangers are lurking (Kahneman 2011):

- In general the law of least effort applies even when the resources of the brain are —whatever is possible—delegated to System 1
- The attention represents a limited resource and cognitive ease is preferred

- Intuition is an essential resource of our cognitive system and can be influenced
- The subconscious mind acts as an associative engine that operates on the basis of pattern recognition
- Cognitive relaxation promotes creativity, whereas blocks of the mental resources during cognitive stress impede good decision
- Confidence, Overconfidence, trust and "priming" and a misleading frame in the decision making ("framing") are common causes of errors even made under conscious control (System 2). That can be avoided by self-critically switching to System 2 and—at best—by modifying System 1 and therefore establishing a learning process.

I am sure some of these points trigger association with an organization you know —not only with single persons and their behavior.

As far as single persons ar concerned, this raises the question of how far this thinking and decision-making processes can be approximately made aware by measurement and feedback and can be made conscious and improvable. One way is to offer so-called "Eye Tracking Glasses". These are new methods for measuring eye movements, for example, when looking at a shelf in the supermarket. A monitor displays in real time whether objects are seen and whether they are also perceived. Mental processes such as concentration and stress can be tracked over time. In test patterns and with test video the ability to focus on surprising details is tested. At the same time the users of such tools learn that "to see" is not equal as "to perceive". In the scenario of an accident you probably hear the phrase "I looked, but I did not see". It is an expression, that though he sees traffic signals, he has not realized that link just as being relevant for behavior. In this case he was distracted and not focused. Also the human aging process can alter perception and attention.

The analysis of facts and data is characterized by the ability to create statistics and graphs and interpret them. The creation of representations and statistics is also often a given opportunity to manipulate something in your own sense. Often subtle methods are used, which are not obvious, clumsy attempts to manipulate. Here a great deal of attention should be applied. The analysis is, as Kahneman says, not trivial, but an error-prone activity, since many potential pitfalls of System 1 are threatening the analysis. Often biases and prejudices, imprints ("priming"), contextual references ("framing") etc. influence analysis unconsciously. The saying "do not trust statistics that you did not create yourself" has a core of truth.

A lately arising new term in this context is "Neuro Business" (see Brown 2013). While we have learned to behave logical and rational in our daily professional decision making in our organization, we should also be aware of how much the human irrational momentum controls us in professional decisions. We have something like an "action mindware", which is marked by our life experience and shapes our motivation in the decision situation. Accordingly, our perception filters are set. Behind the constellations of our value systems are the personal attitudes and the beliefs of the people involved (here not the faith in the religious sense is meant). Ultimately, it is—still largely until today—always the humans who make the decisions.

Most of these patterns of behavior are also applicable on organizations where humans are embedded in a set of structural, organizational and technological settings like Artificial Intelligence etc. thus forming a "subconscious mind of organizations" which interacts with the conscious activities performed by the organization or organizational units. You easily can imagine, that it is possible to design and influence the subconscious mind of the organizational units or organizations as a whole.

2.9 Coping with Uncertainty

This book is mainly about decision making. We try to predict the future as best as possible and new technologies support this with tremenduous progress e.g. by enabling algorithms that support simulation, prediction, machine learning etc. Digital analysis is capable of making things and connections visible which otherwise would remain invisible. Control of processes, individuals, organizations etc. resides in knowing what is likely to happen in the future. Personalisation has become the new normality—both are about control. Taming of chance by combined probability theory with big data has progressed, but the individual remains a random variable. "Predictive analysis und performativity of big data have become possible because we succeeded linking the individual with the aggregate level" Eva Nowotny says in her book "The Cunning of uncertainty" (Nowotny 2015)

Risk management proliferated by transforming the unknown into something known by converting danger into a risk that could be calculated and hence contained. Risk is conceived as evaluation of a potential loss. It has become mandatory to anticipate, assess and wherever possible integrate risk management in the development of technology. Risk has become a known unknown, which one would like to know.

Regulatory systems everywhere increasingly rely upon the control of control—self-checking, self-reporting arrangements in line with performance objectives, measurement and monitoring. Based on the feedback their performance objectives are iteratively adjusted anew—they follow the logic of performativity: a model validating itself in the sense of making itself successful. The borderline to the phenomenon of the self-fulfilling prophecy is a slim one.

"The real problem is that technological risk often ignores social risk" Nowotny says. "Data are the thin description of the self; the perception, self-awareness, self-confidence and doubts are encoded in the thick description of the self". In the past identities were formed and transformed through beaureaucratic processes like in communism. Now there are different political processes on their way like autocratic regimes that try to control everything by using technology, that promised endless diversity and freedom a few years ago. Think of the "arab spring". Today we know that facebook, Google etc. can also be used—with their underlying and self-optimizing algorithms—to create and control perception as well as cognition of information consumers of the relevant target groups by encapsulating them in a "filter bubble" or "echoe chamber", where peergroups or political parties (esp. populist parties) can establish some sort of "social bubbles" that seem to be sustainable in the sense of their creators. These are simply using the inherent mechanisms of social media and exploiting the low skill level of the users concerning how to critically deal with information overflow and evaluate the origin, the aim and the quality of the given—or even pushed—information.

"In the end only uncertainty can explain profits and losses", Nowotny remarks. The complexity of the system is the real power base of uncertainty. On the other hand human behavior especially of individuals often escapes predictability.

The onset of extreme events escapes prediction as well—special mechanisms of self-organization can be phase transitions, bifurcations, catastrophes or tipping points. So an individual person as well as organizations tend to be open to decisions following gut-feelings and momentous intuition. That's where the subconscious of the individual as well as the subconscious mind of organizations comes into play. Trying to perceive an organization in its specific environment and its internal and external connectedness in this way will help to manage complexity and handle uncertainty—hopefully resulting in a resilient organization.

Another question is: How relevant is the past for the future? The non-deterministic answer hopefully is: The unexpected is lurking behind the next corner and the odds for tomorrow lie in the future that is radically open—und thus uncertain.

"Complex systems cannot be left to themselves, they require a new kind of ethos arising from a sense of responsibility for being part of the whole", Nowotny says (Nowotny 2015).

Finally mechanisms must be in place that permit errors and allow performance to fluctuate, so that resources can be managed with a sufficient degree of freedom—more you can read in the chapter about how to form the subconscious mind of an organization.

So in the end uncertainty is an important quality of our civilisation we hopefully will never escape. The taming of chance will thus hopefully never succeed. Going back to a main source of uncertainty—to the individual person and its subconscious as a foundation for the metaphoric view on the subconscious mind of organizations —this leads us to decisive elements of the human psyche: Motivation and Intuition.

2.10 Motivation

"Neuro Business" tries—under the "umbrella" of the keyword neuroscience—to put in the center the role which rewards and punishments play in influencing the subconscious mind of the people involved and the role that motivation plays in this context. The correct approach is that it can be assumed that the brain and especially our subconscious in its "programming" does not distinguish between business and personal decisions. This should also be aware in all decisions concerning staff selection. Nobody has the same "mindware" as his colleague. Everyone forms his reality accordingly (see Brown 2013). From the wide variety of possible realities we choose our reality based on what is familiar to us and familiar to what seems convenient and easy to us and what we expect. We form our reality in a three-step process of displacement, distortion and generalization.

Sprenger (2013) examines motivation fundamentally and dispels many myths—especially the value of the extrinsic motivation. He contradicts the widespread credo that the motivation of the employees is an important task of the manager. Motivation actually has a high impact on the success. Motivation of people depends to a high degree on communication. The extrinsic motivation through incentives and—"false", in the context of motivation often "seductive"—communication cannot and must not replace or even try to replace the intrinsic motivation. Sprenger makes the point by saying: "Instead of moving people in an endless loop of incentives to the desired behavior, he should be better taken seriously; to perceive him in his being as he is and to communicate your expectations on him". "Do not seduce—challenge him" should be the motto. The will to perform is given to all humans. Recognition for performance in family and social—and therefore also organizational—environments are key motivators. The dimensions of performance are the willingness to perform, the ability to perform, competence and the opportunity to perform. Motivation for performance is the responsibility of the employee and not of the manager of the employee. The ability to perform can be raised by education, training and the granting of opportunities to gather experience, partially promoted by executives. The recruitment of personnel has a major impact here. It is key to offer opportunities to perform and that means also to urge performance and agree upon that. Goal setting and the associated process of consensus building are essential and not so much the goals themselves. They are a basis for ensuring that the willingness, the ability and the opportunity to perform are given and that the decisive motivation—self-motivation—arises. Missing opportunity to perform over a prolonged period also destroys the motivation and self-motivation. Performance is not something absolute, but more a matter of expectation.

How these aspects of leadership, like motivation in an organization with its methods, traditions and "culture" are dealt with, and how they are in interaction with the purpose, the vision and the objectives of the organization, forms an essential part of the subconscious mind of organizations. That will be worked on in detail in this book in Chap. 5 where the "shaping" of the subconscious mind of organizations is elaborated.

2.11 Intuition

Few terms in psychology, cognitive science, and also in the considerations for decision making in general and decision-making under stress in particular is so close to the subconscious mind and is so closely associated with the subconscious mind, as "intuition". This term and this phenomenon "intuition" is also highly relevant in an analogy as the "subconscious mind of organizations". From a special behavioral psychology perspective, it was also considered in the reflections on "Thinking Fast and Slow" according to Kahneman (2011) in describing System 1 and System 2.

In the psychology of Carl Gustav Jung intuition is a basic psychological function that allows a perception of future developments with all its options and potential. It is mostly perceived as instinctive detecting or as emotional idea or hunch. The concrete intuition mediates perceptions that affect the reality of things. However, the abstract intuition mediates perceptions in idealized contexts. With Jung's intuitive type of character, a fusion with the collective unconscious often emerges (see Jung 1995).

Understood as basic human skill, intuition is the key ability to process information and to respond appropriately to great complexity of the information that has to be processed. This very often leads to correct or good results. There are two different levels of intuition when decisions are taken: The decision based on emotion and feeling and the decision based on an intuitive mind (incubation). The information is processed unconsciously and awareness is "turned on" when the subconscious encounters a solution. Intuition does not necessarily mean an immediate solution; often it helps to sleep a night on it.

Kirsten Volz from the Max Planck Institute for Cognitive Sciences in Leipzig examined intuition using an MRI device (magnetic resonance imaging) (see Spiegel 2007). In Volz experiment she projected incomplete images of everyday objects— each for 400 ms—on the glasses which her 15 probands in the MRI tube were wearing. In some pictures the outlines of the objects were filtered out, so that the objects looked like processed with an ink eraser. When the probands recognized an object, they reported with the touch on a button. From this, the medial orbitofrontal cortex was revealed as "something of a connection device that reviewed the incoming information to see if the brain knows something like this." This region of the brain was more active the less of the original drawing could be seen, because it meant more work for the orbitofrontal cortex. When this region signaled, that there really was an object, another brain structure has been actived—the gyrus fusiform —, which is responsible for object recognition. Only then the probands pressed the button. This division of labor between unconscious and "consciousness" of future activity accelerates the decision because the "gyrus fusiform" thus has access to the "tacit" knowledge, "sunken" in the orbitofrontal cortex.

This fast, pattern-recognizing and constructing brain activity becomes more important the more complex the environment and the more disjointed the information. Without their hidden "tacit" knowledge humans would be hopelessly overwhelmed.

"The mind, that people are using to make wise decisions supposedly is limited, accounting for only a small part of our actual knowledge," says the American intuition researcher Milton Fisher. "Nevertheless, when we have an intuition this is the retrieval of information that we have perceived and stored sometime through the five senses" (Spiegel 2007).

As about 40 sensations simultaneously reach the brain, the constant input is therefore diverted to another memory: the subconscious. "And sometimes from this wealth of knowledge a small scrap of it penetrates the consciousness." "Then we have an intuition", the psychologist Fisher says (Spiegel 2007).

"The intelligence of the unconscious is to resort in any situation to the appropriate rule of thumb", the psychologist Gigerenzer (2007), director at the Max Planck Institute for Human Development in Berlin says. Commonly, especially during difficult decisions, people seem to trust their analytical mind much more than their gut feeling. "Most of us accept that it is unrealistic to assume that you have unlimited knowledge and unlimited time to choose one from many options. On the other hand, we are confident that we would make better decisions without these restrictions and with more logic" Gigerenzer says. "Good intuition ignores information," says Gigerenzer. "Who wants to be intuitive, should allow himself no opportunity to reflect on his actions". Often the opportunity is missing anyway—such as when emergency physicians must care for accident victims and every minute counts (see Gigerenzer 2007).

But on the serious side of life "newbies in a field rather should deliberate and analyze all the possible consequences of their actions thoroughly", recommends sports psychologist Markus Raab of Flensburg University. "Only those who have already gathered experience in a field should rely often on their intuition." "A theory training, in which the players are challenged with many alternatives, does not work", Raab concludes from his experiments. His advice to athletes is: "Collect as much experience as possible" (see Spiegel 2007). From my experience the same applies on professional business and organizational context.

How can we provide information with context to the actual situation of decision —for a single person to decide or for a group or organizational unit to decide? Which conditions should be provided to achieve best possible results with reasonable effort? This will be further elaborated in the chapter on the design of the subconscious mind of organizations.

The problem with the unconscious is that depending on personal experience, the rules of thumb of intuition can also include prejudices. An intuitive guidance of whether our perception is distorted or not, does not exist. That's the point, when intuition can become a pitfall. And that is the reason why people fall particularly easily into this trap, when it comes to theirself and it is not "only" an external affair.

The german chess grandmaster Stefan Kindermann has his experience in the use of intuition in chess. For decisions in complex situations in organizations he created a 7 step decision model—"The Royal Path"—where he covers the range from providing clarity concerning the environment of the decision situation to the generation of creative ideas as well as to forward thinking followed by backwards reflection guided by the key values of the organization (see Kindermann and von Weizsäcker 2010, it will be described in more detail in a later chapter). Hofstadter et al. declared in the book "Die Analogie—das Herz des Denkens" (Hofstadter et al. 2014) (engl. "Analogy—the core of cognition") associations and analogies as an essential feature of intelligence: "Every concept in our thinking owes its existence to a long series of analogies, that have unconsciously emerged over the years, and they have already contributed to the origination of the concept. The analogies have enriched each concept in our thinking throughout its existence. In addition, at every moment of our lives we get our ideas of provoking analogies, that the brain—by trying to tap into using the old and the new and unknown acquaintances—incessantly produces". Analogies are from this perspective also foundation and building blocks for intuition.

Intelligence is the ability to rapidly detect deep analogies and the ability to put our finger on the essence of a situation within a short time. For example, first mathematicians let themselves guide from vague intuitions and insights before. Afterwards they justify their actions and fill the gaps among others with logic (see Hofstadter et al. 2014).

How can we support opportunities for fruitful intuition in the setup of our organizations and the infrastructure of our organizations as building blocks of their subconscious mind? Is it only a matter of the experience and the quality of the individual person involved? Or is it the set of conditions the organization provides?

2.12 Is There a Subconscious Mind of Organizations?

In the light of new technologies our perception and our perception filters are in the process of change from an individual's point of view as well as from an organization's point of view. They thus shape the subconscious mind of an organization in this ever-accelerating business environment we are in.

This subconscious mind of organizations is more than the (conscious and subconscious) behavior of the interconnected individuals of an organization. The infrastructure of the organization—constituted in formal and informal rules, the communication channels and networks, the information systems, etc.—get a new and more powerful meaning. All the more important it is therefore to condition the subconscious decision processes and to build, form and shape this subconscious mind of an organization proactively.

How these aspects from behavioral psychology and cognitive science can be used at the personal level for good decision-making and how the context of the organization—also in the light of an analogy to the subconscious mind of organizations—can be affected positively, is elaborated in the model of the subconscious mind of organizations (Chap. 4), and particularly in the design, building, forming and shaping of the subconscious mind (Chap. 5) before concluding in the considerations of organizations emerging as "Hybrid Intelligences" from the Digital Transformation (Chap. 6).

Beforehand I will describe components of the infrastructure of organizations, including the relevant new technologies in their impact on the design of the subconscious mind of organizations and particularly the relevant information and communication technologies. Dear reader if you are already familiar with these technologies you can go through the following chapter more quickly.

Literature

Bracken BA (ed) (1996) Handbook of self-concept: developmental, social, and clinical considerations. Wiley, New York

Brown R (2013) Human, Wherever We Go, Huffington Post, 12.03.2013, http://www. huffingtonpost.com/rebel-brown/human-wherever-we-go_b_4365274.html

Clark A, Chalmers DJ (1998) The extended mind. Analysis 58: 7–19; In: A. Clark (2008) Supersizing the mind: embodiment, action, and cognitive extension. Oxford University Press, Oxford and New York

Damasio A (2012) Self comes to mind: constructing the conscious brain. Vintage Books Edition

Dijsterhuis A (2006) On Making the right choice: the Deliberation-without-Attention Effect. Science 311

Gigerenzer G (2007) Bauchentscheidungen. Die Intelligenz des Unbewussten und die Macht der Intuition. Bertelsmann, München

Hofstadter D, Sander E, Held S (2014) Die Analogie: Das Herz des Denkens, 1. Auflage, Tropen-Verlag 2014 Verlag C.H. Beck im Internet ISBN 978-3-608-94619-2

Fan S (2016) How to build a mind? This learning theory may hold the answer. Singularity University Blog

Jung CG (1995) Definitionen. In: Werke G (ed) Walter-Verlag, Düsseldorf 1995, Paperback, Sonderausgabe, Band 6, Psychologische Typen. ISBN 3-530-40081-5, S. 474 f., § 754–757

Kahneman D (2011) Thinking fast and slow, Farrar, Straus and Giroux

Kindermann S, von Weizsäcker RK (2010) Der Königsplan – Strategien für ihren Erfolg. Rowohlt Verlag

Kumaran et al. (2016) What learning systems do intelligent agents need? Complementary learning systems theory updated" in trends in cognitive sciences Vol. 20, Issue 7, pp 512–534

McClelland JL, McNaughton BL, O'Reilly RC (1995) Why there are complementary learning systems in the hippocampus and neocortex: insights from the successes and failures of connectionist models of learning and memory. Psychol Review 102(3):419–457 (Review)

Minsky M (1988) The society of mind. Simon and Schuster, New York

Nowotny E (2015) The Cunning of uncertainty. Wiley, London

Spiegel Online, Intuition: Die Macht des Unbewussten, 28 Apr 2007. http://www.spiegel.de/ wissenschaft/mensch/intuition-die-macht-des-unbewussten-a-479900.html

Sprenger RK (2013) An der Freiheit des anderen kommt keiner vorbei. Campus Verlag Frankfurt, New York

Suhler C, Churchland P (2009) Control: conscious and otherwise. Trends in cognitive sciences 13:341–347.

Turing AM (1950) Computing machinery and intelligence. Mind 59(236):433–460.

Chapter 3
Elements of an Organization's Infrastructure—Relevant Existing and Expectable Technologies

Abstract The foundations of organizations such as visions, mission statements, the organizational structure—and corporate strategies as well as strategic objectives derived from them—are the supporting pillars of the core business of the respective organization and they are part of the—in a wider meaning—infrastructure of the organization. These elements and structures influence the behavior of the organization and are part of the subconscious mind of the organization. The new technologies and tools like sensors, internet of things, augmented reality, social media, Big Data, etc. are attached and settled in this infrastructure—they are becoming part of it—and they change the perception of the organization. But they also change the possibilities for interaction both internally and externally. They change the "cognitive processes" of the organization and thus the subconscious mind of the organization. New concepts, such as social collaboration and realtime enterprise accelerate the interaction and business processes and open up new paradigms like the paradigm of sharing—the "shareconomy". Decisions are increasingly being supported by IT (Decision Support) or even automated. Systems and algorithms with learning abilities and Artificial Intelligence emerge or are moving in. How can the organizational and entrepreneurial governance be secured in such disruptive phase shifts? How can the foundations and frameworks of the organization be kept appropriate and stable? How can the subconscious mind of the organization—required as a basis for the fast and efficient action—be formed and shaped in a safe way. How to make it "resilient"? This requires not only communication but also the right algorithms for the self-control of complex systems. However, organizations have to be designed and built so that they can deal with uncertainties and disruptions as best as possible. This requires an appropriate infrastructure—in a wider meaning—as important part of the subconscious mind of an organization and this requires the meaningful use of existing and expectable technologies.

Keywords Organization · Infrastructure · Organizational cognition · Artificial Intelligence · Resilience · Self-control of organizations · Decisin making · Decision support

© Springer International Publishing AG 2017   25
W. Leodolter, *Digital Transformation Shaping the Subconscious Minds of Organizations*, DOI 10.1007/978-3-319-53618-7_3

After the presentation of the behavioral and cognitive psychology and neuroscience as basis for decision-making in the previous chapter, we now highlight the infrastructural and technical aspects, as well as the emerging new technologies that characterize the subconscious mind of organizations and the interface to their conscious mind. They have the potential to shape the subconscious mind as well as the conscious mind of organizations. We interpret the term "infrastructure" in a very comprehensive and including way as you will see.

Then I will show in Chap. 4 a model of the subconscious mind of organizations and explain it with examples in the form of microstories. Based on these fundamentals we will try "to rethink organizations", considering the design possibilities of the "subconscious mind of organizations"—including the infrastructural aspects and the technological possibilities set forth. Finally we will discuss the **digital transformation** as a societal meta-development concerning organizations and—focusing on the collaboration between man and machine—depict a "Big Pictures" of the future of an organization with **hybrid intelligences** incorporated.

3.1 Infrastructures of Organizations

3.1.1 "Soft" Infrastructure

Many organizations already have them. Others, especially some NPOs (non-profit organizations) or SMEs (small and medium-sized enterprises) simply live them. Fighting for sponsoring money and donations some NPOs often have already worded them carefully. I'm talking about values, visions, mission statements and guiding principles.

Good strategic management in "well-run" companies is based on carefully worded visions and strategic objectives. In large companies, this is a widely established practice; in NPOs this is less established and in SMEs these elements of the strategic management are often not explicitly defined, but they are often set and lived authentically by the business or company owner.

Is a strategy and/or are principles, values and visions already an infrastructure? I believe yes. By definition, infrastructure is indeed always associated with technical aids. But methodological tools as well as the organizational structure are among the foundations that make it possible to perform the company's core business and core activities of the organization in general. It is this "web" of predominantly methodological infrastructure elements that permits an organization to thrive. So let us describe this as "soft" infrastructure.

Mission statements, values, visions and resulting corporate strategies and strategic objectives are essential elements of a good alignment, focus and control of a company or organization. Good planning and controlling systems and processes are also an indispensable part of a management system. Distinctive parts of strategies and management systems, such as information management strategies, environmental strategies, quality management etc. are of great importance for a holistic orientation of the organization.

Even today, many companies are structured and organized very hierarchically. Project management and process management often struggle to break up the silo structure of the respective organizations. Excessive hierarchical organization often is the reason why sometimes the power is lacking for the achievement of key objectives and expected innovations. In many cases, this breaking of the silo structure succeeds under the pressure of competition in the respective market. But as long as the remuneration and incentive systems overly emphasize the hierarchical component and the hierarchical responsibility—resulting in very high, sometimes excessive remunerations for upper management—inevitable tensions and stress fractures can lead to the exodus or—at least—the inner emigration of good and valuable employees.

Often it is attempted to support the ethical direction of the organization and the staff with a code of conduct for the organization. Some also try to manifest the embedding of their organization in society as a whole—going beyond the target costumers and the market for the organization itself—in a CSR program (Corporate Social Responsibility). The supervisory bodies and the owners of an organization are increasingly trying to ensure internal control systems by implementing additional external control systems, e.g., courts of audit in public organizations, auditors in private-sector organizations etc.). They are supposed to continuously monitor the risk management systems in companies and organizations concerning their effectiveness. All these subsystems of a management system can be seen—in a wider sense—as part of an organization's infrastructure, because they are trying to control the company's business and to evaluate if the tasks in an organization perform in an orderly manner. Thus all these elements try to facilitate and to support the organization in order to make it sustainable and future-proof.

When the management system with its subsystems are overly dimensioned, this usually leads to bureaucratic necessities for the employees of the core business. They might experience this as expressed lack of confidence in them as employees. This mistrust clearly might be sometimes justified. Whether it is wise to—under clear assessment of the risks—eliminate some elements of this "mistrust organization", has to be discussed and decided in the respective organizations. This clearly is a matter of the values of the organization. The future economic and social developments and especially coping with the digital transformation are increasingly demanding agile, flexible and adaptive organizations. It seems that in this perspective it would be appropriate for many organizations "to think again" and adjust their control systems.

The theoretically positive interaction of visions and goals with the management systems and the actual practice often have little to do with each other. The motivational structures and cooperation structures of employees, of the management practice, etc. make up the corporate culture. If you want to "think organizations new", it is necessary to understand the informal structures and processes and the actual corporate culture as well as possible.

An essential characteristic of an organization—which in turn is determined by its infrastructure and its organization—is the learning ability of an organization.

Peter Senge in "The fifth discipline, the art and practice of the learning organization" (Senge 1990) identified the cause for why many companies have a much shorter lifetime than humans and many companies disappear before the fortieth year of their existence: It is the inability of the companies to learn. The primary threats to the survival of businesses do not come from events that will easily find our attention, but of creeping changes in the "underground" for which we are to 90% blind. This not surprisingly coincides with the often made claim that the unconscious accounts for 90% of our mental resources. The comparison with the iceberg whose mass is about 90% under water, is indeed made repeatedly.

Peter Senge identifies and models system archetypes that make it easier for us to essentially understand our organizations in their non-linear behavior. This systematic thinking helps us to identify the trends, opportunities and threats, when it comes to understand the infrastructure and the subconscious mind of organizations and to make them targeted. The technological possibilities such as decision support systems (DSS)—some then in daily practice develop into "decision automats", when decision makers confirm the suggestions of the systems without reflection— change the timing and behavior of these system archetypes sometimes considerably. It appears worth trying to rethink organizations in the metaphoric view of the "subconscious mind" as analogon and as framework and "model". Doing that we have to take into account the new technologies concerning augmented perception, decision support and automated action as well as their impact on the performance. Peter Senge's thinking is anchored in the attitude of an holistic view ("holism") and in the multimodal linking ("Interconnectedness"). He says that the basics of systems thinking are best provided at the level of principles of and for an organization ("guiding principles and insight into basics"). These principles have influence on the structure as well as on the behavior ("structure influences behavior"). Complex systems tend to resist any attempts to change their behavior ("policy resistance"). Therefore, it is important to find the right starting point for leveraging and the right lever to implement change. At the level of actual practice (What is happening? What to do?) the system archetypes and the simulation of the system behavior in a specific constellation of the acting system are significantly helpful (see also the section about learning and adaptive systems).

3.1.2 Technical Infrastructure

The technical means of infrastructures of organizations—physical infrastructures— include for example the area of Facility Management. Besides the fundamental importance for the daily operation of the organization in the form of the provision of adequate working space, meeting rooms and other conditions for working, this is of particular importance in shaping the capability of an organizations to provide sufficient flexibility. It has to be easy for the organization to have access to purposeful work rooms and technical facilities according to the employees' actual organizational needs. Especially the direct cooperation between employees with its verbal and nonverbal communication still has to be facilitated in the future thus also enabling familiarity and conviviality between collaborating team members etc.—in spite and in addition to all possibilities of telepresence (from videoconferencing to telepresence robots).

If a person does not have an adequate work space or station, access to proper gear etc., not only they might decide to take one behavior instead of another as a means to "save" hassle or problems, but also, the concentration, focus and efficacy of the person might be affected and, hence their decisions and actions might be different. This can happen at a supervisor or manager level (having also a domino/ripple effect through the organization) but also at worker/production level.

These infrastructures will also be of special importance in times of crisis, for example, if public infrastructures fail due to blackouts of power grids or in case there are massive and continuing difficulties in the communication infrastructures. In such phases the technical infrastructure is an essential factor for the resilience of organizations (Ungericht and Wiesner 2011).

3.1.3 ICT Infrastructure

In the much more interconnected world of today and the future, the ICT infrastructure as a specific element of technical infrastructure plays an important role—not only, but also in terms of resilience. The ICT infrastructure including application programs, such as ERP programs (Enterprise resource planning), computer aided design and engineering (CAD, CAE), manufacturing execution systems (MES), customer relationship management systems (CRM), supply chain management systems (SCM), Human Resource management systems (HR), business intelligence systems (BI) etc. as well as the office automation and workflow management systems are subject to special changes concerning the requirements they have to meet.

Networks including telephone services as well as storage and server systems are increasingly becoming a commodity, and increasingly "disappear" in the cloud. Particularly mobile devices such as smartphones, tablets and laptops with the "user experience" shaped for and by the consumer market as well as the app architecture drive innovation for the above mentioned application systems such as ERP, CRM etc.

Then there is the advent and the increasing use of social media in organizations and companies. They are often opinion-forming on the one hand, and on the other hand they form communication and distribution channels for the organizations towards the end user and to the market (Business to Customer—B2C). The business to business communications (B2B) via portals and collaboration platforms—often temporarily or limited to a project—enable some sort of virtual enterprises, e.g., in collaborative engineering and project planning and execution. An early form of virtual enterprises were the consortia and alliances formed in the film industry, organized by the producer of the respective film ("Model Hollywood").

The Big Data technologies that are currently being tested and are partly already operated by numerous companies are just in a hype phase, in which there are almost no limits to imagination. They have the potential—and they already do-, along with the Internet of Things (IoT, Web 3.0) and the possibilities of the machine-to-machine (M2M) communication to trigger a disruptive wave of innovation in many areas—in service industries, in facility management, in production (industry 4.0), in healthcare etc. establishing new products with embedded systems and connection to the Internet. Decision-making processes are changed by the possibilities of predictive analysis based on Big Data. These predictions are projections in the usually near future or—in control systems with their high degree of automation—immediate future based on algorithms and their underlying models. These models in turn may by appropriate algorithms from the "extended perception" (by means of comprehensive and networked sensors), which themselves constantly improve (machine learning algorithms). Klausnitzer even speaks boldly about the "end of coincidence" (Klausnitzer 2013) when predictive technologies succeed in many areas of life (*I personally am convinced that uncertainty will stay a predominant force for future developments—at least as long as the humans stay in the drivers seat and AI will not take over, something I would not expect for the next decades–contrary to Ray Kurzweil's predictions concerning the singularity expected 2045* (Kurzweil 2005)—*see below*).

All this will bring up new products and new services, change business processes, give rise to completely new business models, drive out other business models from the market, create new companies and bring other companies and organizations to disappear. The digital transformation is happening—with or without the affected organizations. Experts speak of the emerging "real time enterprise" referring to the future with its ubiquitous sensors, the big data technologies and the numerous mobile devices up to the "wearable computer" (the bracelet, the watch, the wearable sensors), and the possibilities of the extended perception (augmented reality with the Google glasses as prominent example), but also with other much more profound applications in surgery, maintenance, cyberwar with remote controlled weapon systems, etc. The way leads to self-regulating and self-learning autonomous systems we already see in robots—with concepts, ideas and sometimes terrifying scenarios beyond—from massive Artificial Intelligence to transhumanism. Every organization will have to face that and actively have to work on that digital transformation. The metaphor and framework of the "subconscious mind of organizations" as a "model of thought" maybe might be helpful to meet these challenges.

The pressure for change and adaptation on organizations will be tremendous when these technologies and the "real time enterprise" gradually grab place. The responsiveness of the organizations is required to speed up. Companies with the agility and controllability of the often-quoted tankers will have to organize differently. Proactive design is required.

The design principles should "consult" successful living systems—notably the "crown of creation", the human and our vital organ, the brain. In technology this "consulting the nature" is already quite common: Look at successful examples in bionics.

Ray Kurzweil, a pioneer of Artificial Intelligence and the father of speech recognition in his book "How to create a mind" (Kurzweil 2012) already has gone far in his approach of reengineering the brain. In his book "the singularity is near", he describes and predicts the singularity (Kurzweil 2005), when systems of Artificial Intelligence will be able to reproduce themselves, to develop themselves and organize themselves—setting their own goals and finally their own agenda. I am confident that this will not take place and the governance of human intelligence will remain intact, as well as the organizational capacity of people—led by hopefully intact value systems accompanied by good political governance. The unpredictability of humans will always result in uncertainty—hopefully in the right direction.

It therefore seems reasonable to seek analogies to the structure of the brain and other features in living systems in the future design of the infrastructure of organizations—particularly in the design of the subconscious mind and its interfaces to the awareness and the conscious mind of organizations. Analogies are—following Hofstadter et al. (2014) the core of cognition—the "heart of thinking". It seems obvious to try to apply findings of cognitive science, psychology and behavioral psychology not only to those persons working in business, their collaboration, their way to handle and solve conflicts etc. but also to apply it to the organization as a whole: Let's try to proactively shape the conscious mind—which seems easy—but also and especially the subconscious mind of organizations. For a successful "Real Time Enterprise" a clever designed and well-trained subconscious mind ("System 1" of an organization—following Kahneman 2011) embedded in good governance ("System 2") will be indispensable in the future. The infrastructure of the organizations—in this broader meaning described above—must be prepared for it. Let us think the organizations new—let us design, build and shape the organizations of the future and enable them to be agile enough and sustainable enough to persist. We will need it in view of the predicted age of Artificial Intelligence. As we see in this 360 degree view, infrastructures are an essential part of what is being developed in this book as the "subconscious mind of an organization".

So let us next consider some of the relevant technologies in a depth that is appropriate to the task of this book and relevant to the organizational structure of organizations, to the organization of their subconscious mind and to the interface to the conscious mind and the awareness of the organizations.

3.2 Relevant Technologies and Megatrends

Now I will briefly describe some interesting technologies relevant to the design of the "subconscious mind of organizations". They cover the issues that influence the cognitive process from perception to recognition to decision and to action of the individual employee as well as—metaphorically spoken—the perception and cognition of organizational units and of the organizations as a whole. The explanation will be made without dipping too deep into the technologies, it will describe relevant developments in a way that should be suitable for the non-technical reader too. The—in new technologies—technically experienced reader can go through it in a "fast track", but he should keep in mind how these technologies affect perception and cognition of both individuals and organizations. Concerning the key technologies in AI (Artificial Intelligence) the Sect. 3.2.15 also includes some philosophical and societal considerations.

First, a brief sketch of the development of these technologies:

3.2.1 Developments in Information and Communication Technologies (ICT) and Their Application

After the first computers which were still working with vacuum tubes a rapid development began with the development of the transistor. In the 1960s and 1970s, the mainframe architecture dominated (mainframe with data and program code on one computer)—initially controlled by punch cards and then controlled with simple terminals without graphic capabilities etc. The dominator on the market at this time was IBM.

From the mid-1970s,—along with the increasing miniaturization of the electronics in particular with the development of integrated circuits—intelligent workstations primarily for engineering tasks and first minicomputers in conjunction with these workstations were organized in local networks. Now data and program code were already partly on separate computers. The ability to distribute the "intelligence" created new opportunities. In this era of the "minicomputer" these new devices coexisted with the mainframes and initially focused on niches (for example engineering). Then they went into competition with the mainframe after increasingly commercial applications became available on those "minicomputers".

In the 1980s, the personal computer, the PC, appeared with the pioneer Apple and the "follower" Microsoft, which soon established itself in the dominant role. With this now available PC technology, which was increasingly able to take over tasks that had previously been reserved only to expensive workstations, the architecture of the "client-server computing" became established—further emphasizing a distributed architecture. The applications—and thus the majority of the program code—were now running on the client or on "application servers". The data were still held more centrally on so-called database servers due to the integration requirements, which were derived from the need of continuous business processes.

Parallelly the Internet evolved in the 1980s—with the breakthrough in the 1990s due to the growing bandwidth available on the public and private networks. The Internet became the "nervous system" of organizations and of the society. The mainframe architecture had also evolved in parallel pathways—claiming its position especially in large companies in the financial industry and major research institutions. Thus, beginning from the mid-1990s an ubiquitous "web" or "woven fabric" emerged out of the network itself and from networked systems with all their intelligent terminals, PCs, servers and mainframes.

Questions of user experience were mainly focused on professional users then—but homecare-users also used these interfaces already successfully (Windows being the major product). New communication technologies such as Bluetooth, WiFi, RFID and NFC now allowed mobile and local networking in a quality that the user was more and more "consumer of a service" and not any more "operator of a system". Many successful professional systems and applications with a long history of development like SAP etc. were and still are confronted with "consumers of services" instead of "users operating an application". This new attitude puts considerable pressure on the providers of these long standing "legacy systems" especially concerning the required user experience. The information and communication technologies are embedded in an increasing number of devices, systems and products—so-called "embedded systems"—, a fact that is increasingly common for people but is hardly perceived consciously.

The sensor and actuator technologies—combined they are called mechatronic systems—first came to use in special isolated cases, for example in the form of thermostats, pacemakers etc. These systems can now—with the new communication technologies such as NFC (near field communication) and "mobile to internet" be incorporated and integrated in the "web" in a controllable way. The Internet of Things (often also called Web 3.0) arises. Thus, the possibilities of biometrics and the increasingly penetrating standards for interaction of systems and their interoperability from machine to machine (M2M) and human to machine become more efficient and proliferate.

Process integration across organizational boundaries is highly automatable and feasible in high quality and reliability. This results in highly interlinked systems, which are described with many new buzzwords: ubiquitous, pervasive, ambient computing (ubiquitous, penetrating, adapted to the environment).

In the private sector, initially the network was—next to the email communication—primarily used as a search engine for finding information (Web 1.0). The user was sort of traditional consumer.

The generation of "digital natives" has realized these powerful techniques as a new form of interactive communication with all the possibilities of self-expression as well as the interaction with other interested or relevant participants. So in the first decade of this century Web 2.0 emerged exponentially. In this area Facebook has indeed established a very dominant role—some say a monopoly, though some new competitors like Instagram etc. already established themselves.

The increased use of mobile devices, the convergence of telephony and information technology in the user interfaces as we know them today from smartphones and tablet computers where again pioneered by Apple. Now almost all organizations, businesses and private users of all generations are merged in this ubiquitous network.

As mentioned before, the data and the programs were localized on the mainframe—on one computer—where the data were on drives and therefore not directly accessible for the processors (and their associated fast access memories) where the programs were executed—sometimes distributed over the network. Now the necessary IT resources are offered as a service from the "cloud". That sounds and—partly is—somewhat "cloudy" one might ironically remark. Often this is beyond comprehensiveness for the majority of people. It resembles our attitude towards electricity: "electricity comes out from the socket". Cloud computing takes place in the private cloud—the enclosed area (for example, in a safe company network)—, in the public cloud (for example Dropbox, publicly available from everywhere, but password protected) or in mixed configurations (hybrid cloud).

Meanwhile, there are powerful computers with the ability for massively parallel executing of program code in parallel processors—either on the same computer or distributed in the—now already—very fast Internet. But with the new technologies the data access does no longer sufficiently perform on the "slow" spinning disks, but the processors have direct access to huge memories with terabytes of data—the so-called "in-memory computing". This huge amount of data can be processed quickly with new algorithms, enabling complex analysis and visualization in a "realtime" setting. It is thus possible, to have a "dialogue" with huge amounts of data and find new and unexpected insights and discoveries and even to make predictions in high quality—combined with appropriate visualization based on algorithms and optimized self-learning models in the background. Keywords in this context are Big Data, Visual Analytics, Predictive Analysis, Knowledge Discovery, Deep Learning etc. It is easy to see that these developments open completely new perspectives towards Artificial Intelligence (AI).

IBM pioneered this area with the system "Watson" and made AI evident to the public. Watson won the popular quiz show Jeopardy, makes predictions and gives advice for physicians in the field of oncology "drawing knowledge" from reading numerous medical journals. A predecessor of Watson was the chess computer BigBlue from IBM. BigBlue defeated the world champion in chess. AlphaGo—developed by Google has defeated the world champion in playing Go in 2015—years ahead of the prediction that this might succeed.

In the robotics area the increasingly sophisticated tools of Artificial Intelligence merge with advanced features and abilities of sensor technology and mechanics, as well as information and communication technologies. Formerly industrial robots such as welding robots and assembly robots accomplished quite complex but recurring automation tasks in production and were the vanguards of technological development. Mostly they were "fenced in" due to security aspects.

Today, robots in warehouses (e.g., AGV's = automatic guided vehicles), minesweeper robots and drones in the military, and industrial robots are state of the art. Telepresence robots, telemedicine robots, surgical robots and robot nurses are expected to be widely used in near future moving along with humans like work assistants and companions. Even in households robots are already widely used for cleaning, mowing the lawn etc.

Closely related to the area of Robotics in development, programming and control are the subjects of virtual reality (VR) and augmented reality (AR). In connection with the advanced sensor technology they enable humans to expand their tools and even manual actions over great distances. Drivers in these developments were the (computer) gaming industry and the film industry with their animated and science fiction films.

Thus new opportunities for collaboration across organizational boundaries develop—within organizations, between organizations, from organizations to customers, clients, patients, and other stakeholders. Keywords in this context are: "shareconomy", "social collaboration" etc. Google with its "near-monopoly of information" in the form of its search engine, Amazon in the field of the book market including ebooks and increasingly expanding to other industries, eBay, etc. take advantage of these developments in order to attract the attention of people for their services—without generating content.

Content providers—e.g., authors who want to publish—get less for their work. Google and the other "market makers" thus also conquer the advertising market. The "economy of attention" is thus threatening to wipe out quite a bit of variety and diversity. The Digital economy—as a generic term—leaves no stone unturned. It generates new potential risks for previously successful business models and entire industries, but the digital economy also generates potential risks for society and democracy. Just think of privacy, confidentiality, data protection, manipulation in "filter bubbles" of social media, disaster security as well as the resilience of our increasingly "digital civilization" in general.

The opportunities of digitization thus provide new tools for people and organizations. These tools will soon be used by trained humans partly subconsciously—remember the role of the subconscious after you have learned and practiced driving a car. The possibilities for people to network within organizations and from the organizations outwards can be considered as unique in the recent evolution of mankind. They are already extensively used. The pace of these developments was and still is enormous. Now not only the people but also the organizations get new tools on hand that go far beyond communication and networking, tools that might be adopted as fast as never before and tools that will become immersed in our societies as well. So it makes absolutely sense to speak of the subconscious mind of the organizations and to analyse this subconscious mind and its interface with the conscious mind and the awareness of the organization in order to shape the way this subconscious mind works.

Now a brief explanation of individual relevant technological developments and issues for the topics of this book:

3.2.2 Network Technologies

When we apply the analogies of the brain and the nervous system to the network technologies and their development, it is evident that the connecting structures in the brain and the central nervous system require enormous capacities to perceive what is going on and to accomplish what we and—concerning our subconscious —"our brain wants to happen". It requires massive parallel processing of information and communication.

At least the massive advances in technologies concerning the provision of bandwidth at a sensible cost-benefit ratio have enabled the rapid development of technology in the information and communication technologies. This applies to the fixed network, in which the fiber-based technologies have allowed a quantum leap —coupled with enormous advances in the algorithms and the software in "packet switching". At the same time—and building upon this fiber based technologies as backbone—the mobile sector and its bandwidth have developed dynamically. Thus access to the network has opened for very different sections of the population. In the health sector for example—according to American experts (Medical informatics world conference 2014)—especially for the poorer classes smartphones and the mobile networks allow access to health information, prevention and structured interaction with health service providers. Those poor people usually don't have access from PC's at home.

Telephone and Internet are increasingly using this same infrastructure—not only in the public networks but also in the corporate area. That leads to further dependencies and other risks associated with disturbances. Resilience of these networks is a key question concerning the resilience of our today's society. "Blackout" and "Offline" are increasingly discussed. Risk mitigation is on the agenda. Our experiences with massive blackouts are—thank god—still lacking. If one believes in the LOAR (Law of accelerated return) by Ray Kurzweil (Kurzweil 2012), there will be a more dynamic development in the interaction of communication, computing and storage technology due to the parallel exponential technological advances in each separate field. Kurzweil is trying to develop Artificial Intelligence by the way of reengineering the human brain "in silico". Finally he (and his successors) already succeeded in the field of speech recognition—though already starting at the beginning of the 90's.

While networks as we know them today in the professional use are strongly focusing on business-to-business use (B2B), we are about to see new networks, which are networks of relationships for and with customers, or special networks for financial services with its own security structures. The network effect—more participants make the network more attractive and increase the value for all—is only effective if the networks are open and can grow rapidly.

If you compare the network topologies of the brain with the evolving network topologies of the Internet and corporate networks in an analogy, it becomes clear that the Internet in its evolution has allowed on the one hand very distributed structures concerning the content (in contrary to the central role of the human brain). However the Internet allows the concentration on a few "world brain regions" concerning the linking of content, what is more central like the human brain on the other hand. At this point the information concentration at Google, the concentration of social media with Facebook and the concentration of e-commerce with Amazon, eBay and AliBaba have to be mentioned—though they have within themselves distributed structures again. At least these technological structures favor the tremendous aggregation of power up to some form of monopolies.

Let us "expand" the view: The human nervous system and the brain end with the death of each human. From Internet and communications infrastructures we expect more: "They are not supposed to die." Though they have yet to prove that, as they are still in the middle of their life span, when you compare with a human life span. The Internet and its adjacent technologies have become a critical resource in terms of the continued existence of our civilization and are an essential part of the infrastructure and the subconscious mind of organizations. The resilience and sustainability of our organizations are massively and critically dependent from the operation of these network technologies.

Knowledge and experience in these infrastructures are not available so highly redundant as the distribution of knowledge on nearly all humans as far as normal everyday life is concerned. So it is important to be cautious for the sake of our organizations and our society. If—as it is increasingly predicted—the power grids rely on the Internet, as it might be expected in view of the development of smart grids, and the Internet does not work without reliable power supply, so this is a matter for concern. The book Offline! (Grüter 2013) and the novel Blackout (Elsberg 2012) give a good insight in this regard. Organizations have to secure a certain time span (a few days) with emergency power supply etc. before precipitating—critical infrastructures are supposed to have more emergency power supply. But: in case of longer blackouts, when the power grids lack fully available Internet and can no longer be booted, serious damage to our civilization threatens.

Technological ICT platforms grow closer together (fixed net, cellphone net, cable operators, TV service, video offering). The Internet will be flooded with many additional terminals and communication partners, if you consider developments like the Internet of Things and the Internet of Everything. But who considers and regulates the necessary decoupling of systems and system elements in order to come to proper resilient topologies? Will the market do this? Will the market regulate itself? How will the market behave in the first major blackouts? Who will be responsible? Today we see government authorities becoming weaker and weaker —compared to the market. Who regulates this transnationally and internationally?

3.2.3 Internet of Things, Internet of Everything, Embedded Systems, Sensor Networks, Mechatronics

The Internet of Things, according to the Gartner Hype Cycle was considered to be at the top of its hype in 2014, and a descent into the "valley of tears" was imminent —before it would then continuously reach the plateau of sustainable use. There is no doubt that the interconnection of different devices such as sensors, mechatronic components etc. will rise dramatically. These elements will be rapidly connected either individually or in the form of embedded systems and wireless embedded systems. As the standardization processes in this area are still in progress and the broad and efficient use requires accordingly affordable prices, it is necessary to accelerate the standardisation process. However, the optimistic forecasts will probably not become true at such short notice.

The standardization process might accelerate especially as the technological trend for "software-defined anything" is rapidly progressing (Gartner 2013). This means that the infrastructure becomes increasingly programmable and various devices using standards can relatively easily be integrated (similar to—albeit much more versatile and powerful—the programmable logic controllers in the automation wave in the 80s and 90s of the last century). The interoperability to databases and thus the ability to integrate larger systems by innovative use of data (Big Data from the IoT) will be enabled by cloud technologies and their inherent virtualization features.

This may have a massive impact on products and business processes themselves —as the stories in the following chapters about the maintenance of a rolling mill in a speciality steel group clearly show.

Imagine that the energy management of buildings incl. the shading systems of glass facades, the control system of a sensitive plant or the heating control functions in homes are embedded in the vastness of the Internet. Considering available technologies coupled with the potentials of cybercrime or simply the spreading of new computer viruses, it is evident that we urgently have to establish decoupled structures with redundant control functions and a high level of safety and compliance in order to avoid severe problems. How will we deal with it in case of massive malfunction and disturbance? We will see, whether there will be enough qualified and trained personnel for an offline or manual mode in case of failure and whether there will be enough resources for the necessary recovery process—think of the ever-increasing pressure on productivity and efficiency.

Networks of sensors, actuators, standalone or configured for a particular purpose in form of "embedded systems" can collectively be characterized as the Internet of Things. They are thus part of the infrastructure and the subconscious mind of an organization: They quickly work in the background, efficiently and reliably as the "system 1" following Kahneman, so that we do not perceive it consciously—the organization perceives it subconsciously. It remains to be seen, whether we can take action efficiently and regulate and control the system when e.g., in case of failure, the "system 2"—the skilled personnel—has to be addressed, which is then possibly outsourced to infrastructure and ICT providers.

In case of failures or criminal attacks that affect many organizations, the outsourcing partners will be called for help simultaneously. The following congestion on the Internet used by all customers and the restricted personnel ressources at the outsourcing partners might well overwhelm them accordingly. We should in any case responsibly prepare for it.

Imagine: all mobile and fixed connected devices, products and services are digitized, including products such as cars, TV sets, refrigerators, etc. This changes the way humans as well as organizations perceive their surroundings and will change it even more in future. The combination of all these data streams and services is almost incomprehensible and therefore unmanageable by humans. To generate models to operate, to constantly develop and expand this and ultimately monetarize it in valid business models is a tremendous challenge for the coming years and decades. The organizational models to orchestrate these connected people, things, information and locations are to be developed from and for each organization specifically—depending on the degree of impact of these developments on the organization. The resulting opportunities and risks still have to be taken into account and accordingly be exploited, avoided or mitigated. All this has to be considered accordingly in the design of the subconscious mind of each organization.

3.2.4 Mobile Devices

A particular class of these infrastructures and systems are the terminals, and especially the mobile terminals that are probably the most important class of system components numerically and otherwise. They are key for most of the interactions of humans with the systems and infrastructures. It is worthwhile to take a closer look at these developments:

For the possibilities of enhanced perception, of personal connectedness and of new ways to interact over long distances for different requirements, there are more and more purposeful different intelligent terminals available that have actually only one thing in common: the SIM card to connect to the network. The user interfaces are depending on form factor and size of the devices—but in its intuitive handling they are similar. As operating system three platforms are emerging (Apple iOS, Google Android and Microsoft Windows), who claim to coordinate the various devices in their platforms. They do this mostly via cloud services for data management, and for the small "software helpers", the apps. In terms of interoperability between these devices and platforms there is still a clear need for improvement.

The classic PC or laptop and—in recent years—the massively popular smartphones and tablet PCs will shape and change user behavior. Smart bracelets, sensors in garments, data eyeglasses, Augmented Reality-equipment with special helmets and "data gloves" will enhance this change. Sensors for acquisition and interpretation of our brain activity to control devices by means of brain activity etc. will mean that our interaction with the environment will change. This variety of devices for personal interaction are introduced in the systems and organizations already partly by the users (BYOD = "Bring your own device"). On the one hand this constitutes an enormous challenge for the security management of the systems and organizations—on the other hand this will enable new forms of collaboration and possibly changes the way organizations work. The subconscious mind of the organizations will change.

It is a big challenge to find the balance between flexibility on the one hand and stability and confidence felt by the user on the other hand. Organizations currently struggle to bring content creation, collaboration and core business applications under one hat—facing the requests for a high variety of devices in the background. This will challenge and/or occupy organizations for some time intensively.

Which devices will prevail? The plurality of terminals will remain and may grow even stronger. The PC, laptop, tablet computer or smartphone will be the control center of the personal network, which includes bracelets or other "wearables" with vital sign functions or the like. For specific tasks headsets and other augmented or virtual reality devices will integrate. These devices have to provide for the secure authentication of their users.

The variety of offerings is large and is even becoming greater. Finally those combinations of devices and the application software which will succeed in the long term will be those ones, which offer the best "user experience" and the best "customer experience".

3.2.5 HCI Human Computer Interface, User Experience, Biofeedback

The number of possible interfaces between man and computer is constantly increasing, but finds itself still largely in the experimental stage. Examples of "state of the art" interfaces that go beyond conventional interfaces (such as Windows and Apple and what the smartphone manufacturers offer) are:

- The recognition of intentions from the user's behavior to the corresponding zooming of image regions (see Biswas et al. 2013).
- Control by will (Brain-Computer Interfacing, BCI) through pattern recognition in electroencyphalograms (EEG) of brain activity. These developments, which were initially designed for severely disabled people, can be used also in "hands-free operational environments" (see Steyrl et al. 2013).
- Voice input is already widely available, for example, the Siri assistant on the iPhone from Apple. Chatbots will develop to companions of their users and will have access to vast ressources of knowledge. The access will be context-sensitive and will enhance the capabilities of the user significantly.
- "Data gloves" to control the three-dimensional space, for example, in the robot programming or with surgery robots.

The essence of the interaction between human and the "virtual world" in the broadest sense—from the accounting application on to the setting of thermostats, heating controls or entire "Smart Homes" and on to the computer games—is the "user experience".

Only systems and work environments make sense and will be of social and economic benefit, to which the user finds access, which are useful for them in work or leisure activities and settings. Such "systems of engagement" that create something like customer and user involvement and participation need the same kind of "user experience"—a common user experience—across multiple platforms with the "cloud" as a framework.

One way to measure user experience are biofeedback methods, such as eye tracking, where the eye movements are recorded and so the flow of attention can be observed and user interfaces etc. can be optimized on this basis. A technical term for this is "usability engineering".

We live in an economy where the value of events and experiences are of more importance than the differentiation on specific product characteristics. Simplified and "straightened" application experiences—such as shopping experiences e.g., teleshopping—are a prime criterion in decisions due to our own internal perception and decision-making processes. That's called the "consumerization" of information technology. Better cooperation processes along the value chain are enabled by appropriate user experience in the use of the systems and thus lead to process- and often product-innovation.

The cognitive load of the user is a growing problem. The top priority in the design of "user experience" must therefore be the immediate support of the actions and decisions in the particular situational context. You can easily imagine how such "user experiences"—if the applications are frequently used—affect the cognitive processes (system 1 and system 2) and in the medium term even affect the sub-conscious of the user. The experiences with young people who spend hours playing computer games or are moving in the social networks for several hours a day are a striking example.

It is easy to understand that such user experiences will also lead to changes in the cooperative structures and cooperative mechanisms within the organizations as well as to the outside. Thus the decision-making behavior changes. The changing influences of "user experiences" must therefore be seen as part of the subconscious mind of organizations—overlapping with the conscious mind and the awareness of the organization.

An essential part of "user experience" is the visualization of situations, environments, etc. The reality is represented digitally. We speak of an extended and virtual reality (augmented and virtual reality—AR and VR).

3.2.6 Virtual Reality and Augmented Reality

According to Gartner (2013) Virtual Reality (VR) and Augmented Reality AR) are in their hype cycle straight on the way down to the "valley of tears". The highlights of the "hype" are clearly exceeded.

Virtual Reality has been specializing on film and game industry for a long time. Another field has been—and still is—engineering, for example the movements of robots can be simulated in virtual reality in order to avoid collisions in real operation. In architecture rooms and spaces are brought to life in this way. IBM presented a wonderful showcase of virtual reality in the 90s of the last century when the reconstruction of the "Frauenkirche" (a church) in Dresden was projected—as part of a sponsorship: Whilst the Frauenkirche was still under reconstruction, it was made virtually accessible. Thus donations were attracted and motivated. Even the use of the remaining stones of the (in 2nd world war) collapsed church was simulated. The Frauenkirche, which today represents one of the main attractions of Dresden, is a good example of how to turn a virtual reality into reality. IBM since then has defeated the chess world champion Kasparov with BigBlue, the chess computer, and has built Watson, which won the quiz show Jeopardy—as further illustrative "beacons" in the development of new technologies. Artificial Intelligence thus has helped to make the current technological developments clearly comprehensible to the general public and thus rendering it more confidence. The virtual reality of the Frauenkirche thus was a first "beacon" for virtual reality (VR).

Let us now go for Augmented Reality (AR). Google Glass as best known example—since many years in the media, has long been only available as test equipment. But it never entered the market. Google Glass still serves as a synonym for augmented reality, where for example the surgeon gets a view on vital sign data of a patient without turning the head away from the patient being treated. Recognition of the location of the user with interfacing geospatial information and available services ("location based services") are also possible with a reasonable user experience. There are already applications—from other specialized manufacturers, not just Google—for example, where the picking activities in warehouses are supported or the construction plans are inserted on the glasses allowing handsfree information when repairing a machine—using voice control to navigate in the information.

VR and AR will partly change the way we perceive the world and what is happening around us. It will change our imagination and bridge long distances. If organizations use this for specific tasks and scenarios e.g., in collaborative settings these new ways to visualize and interact will change how things get done, some decisions will move from "system 2" to "system 1"—thus also shaping the subconscious mind of organizations.

3.2.7 "Social Media" and "Social Business Collaboration"

The social media emerged first from the new technical possibilities and from the need of—mainly young people—to communicate and exchange ideas on issues of their speciofic interest. This gets done in new and different ways—not any more in the established forms of communication and use of media of their parents. With the advent of low-cost cell phones and the ability to send SMS, new communication channels have been created and have been used widely—initially with little content. Anyone who had experienced children on their way to adolescense at the beginning of this century, knows what I mean. Things were notified, photos posted. SchülerVZ (for pupils) and StudiVZ (for students) were first platforms in german speaking countries—predecessors of facebook. Music was "illegally" exchanged via Napster—driving the music industry to despair.

The need for new forms of communication and the sharing of information, the giving of recommendations (the likes) etc. was so clearly manifested by the new generation. Young companies and startups succeeded to form these needs in business models—very profitable business models, as we know today. Larry Page and Sergei Brin with Google, Mark Zuckerberg with Facebook are the best known examples. Yahoo, Twitter, LinkedIn and more developed, some disappeared. Thus a partly new form of communication and economy emerged. Apple was one of the few "classics" from the IT world, which managed to monetarize this new economy —first with the iPod and iTunes as viable new business models. Pioneers like Napster who satisfied the social need of sharing without regard to copyright were sold or disappeared from the market under the pressure of lobbies and lawyers from the music industry. Veterans of the IT industry such as Microsoft could benefit only marginally from these developments, but earned good money in the market for corporate customers as this market continued to perform well.

More and more new models of "shareconomy" or "sharing economy" emerge, in which the joint use of resources and sharing are the main drivers—combining technology and content. This is troubling more and more mature industries in the travel business, car rental, etc. Examples are Uber, AirBnB, Trivago, etc.—the digital transformation is under way.

Many of these new businesses deal with the attention and the time of potential customers—that is the way they monetarize it. The producers of goods, services and content are endangered to become commodity or get squeezed—but they are still providing the jobs. The Internet has opened up the markets and made them transparent. Global companies emerged virtually from zero. Previously this had taken decades in economic history. It is indeed no longer about goods but about the citizens' attention and thus the possibility to give information. These players are monetarizing information via advertising budgets of the producers of services, content and products. In this context it should be noted, that these businesses also managed to establish themselves in the world in a way that only a minimum of taxes are paid no matter where the business is generated. Regulatory policy does not succeed to counteract.

As nearly all activities are covered digitally, the user in the network provide many seemingly worthless data. The value of the user profile and the preferences of the user was soon recognized and used by Facebook, Google etc. to monetarize the "directed attention" of the user. The user became a "target" for placing offers and incentives to him as potential costumers. The fact, that massive conflicts with the subject of data protection (see Sect. 3.2.10) arise, is evident and far from being solved satisfactorily—especially in countries with high sensitivity to the privacy issues like mid-european countries.

This vast amount of data partly seeming to be garbage was the starting point for a new hype that is about to differentiate these data into useful businesses and applications—Big Data (see Sect. 3.2.8). I actually prefer the term of the "innovative use of data" or "smart data".

With the social networks and the associated communication and collaboration platforms completely new business models and cooperation opportunities arise across all forms of organizations. Some characteristics of these options are known as Crowdsourcing.

Crowdsourcing ranges from collaborating on content like Wikipedia to Crowdfinancing—to finance economic activities passing by the established financial markets. There are new opportunities for participation and publication of content and opinions in the form of "citizen journalists", bloggers etc. Many online media operate their own blogs to expand their readership. The Web 2.0 is establishing itself as a medium without hurdles in terms of required capital and licenses such as in broadcasting. It can be used as well for corporate PR (Public Relations) and marketing. Services, for example hotels, hospitals, doctors, etc. are valued by customers, which of course also gives room for manipulations etc. Thus one or another crisis of confidence, leading to disputes etc. will always continue to occur. "Shitstorms" as dynamically evolving—sometimes uncontrolled—campaigns are new phenomena in this new world of attention and opinion that lacks structure on the one hand and provides openness on the other hand. So in Web 2.0. enormous uncontrollable dynamics may arise.

On the other hand an enormous creative and innovative potential lies in it: "Co-creation" is a special form of crowdsourcing, where e.g., corporate tasks are outsourced to freelancers in the form of competitions with the corresponding prize money and bonuses. So supposedly the company Goldcorp has provided geological data for an exploration area on the net, with the result that under such a contest 50 new gold veins were discovered. Collaborative creative competitions are another example. That raises the question whether the collective intelligence of "self-assembled talent" in the net is able to achieve faster and better results than specialized research and development departments. Another example of co-creation is the joint product development with customers. LEGO practiced this in the area of its robot series (see Lego 2014). Motivators for participation are not only bonuses and prize money, but also the attention and recognition of the respective "community".

Sociology speaks of swarm intelligence. Systematics, structure and hierarchy threaten to dissolve. An opinion and control pluralism emerges. Anyone can add something, comment on it etc. Conventional working conditions and guarantees are losing importance or are lost. But what about the quality? There probably are self-regulatory mechanisms in the crowd. The award in such "co-creation" processes is only to those who deliver useful reasonable quality. Jovoto community for handling tenders and competitions for co-creation and crowdsourcing is an example.

From the perspective of the affected organization immense opportunities and threats arise, that can hardly be controlled and managed in detail. Where are the right starting points to this new—more dynamic—environment? How can beneficial, expedient and safe collaborations be ensured in this somehow "amorphous" structure and tight network? The views on what is going on are changing: from the perspective of the person responsible, from the perspective of an organizational unit, from the "surrounding" perspective of an organization or a company? It seems appropriate to understand all these internal and external knowledge sources and the links between them as "subconscious mind of the organization".

The idea is to use this enormous potential by designing the subconscious mind of the organization as a whole and the subconscious mind of the organizational units in detail deliberately. The challenge is to give sufficiently room to the "System 1" (following Kahneman—fast thinking) from the perspective of each hierarchy and its respective responsibility in order to work efficiently and to find good solutions. On the other hand the "system 2" (thinking slow, reflective process) has to be selected and activated at the right time. For example: when does the management from the upper level get involved?

3.2.8 Big Data—Smart Data—Reality Mining—Analytics

A topic which has passed a hype or partially is still in the middle of its hype is big data. What is this? Viktor Mayer-Schönberger, one of the authors of a key book ("Big Data") on this subject (Mayer-Schönberger and Cukier 2013) describes big data as follows: "Big Data is the ability to share information so that new knowledge, goods and services of significant value are created". I tend to describe Big Data as "innovative use of data" and prefer the term "Smart Data".

Data is used as raw material for new insights, new business models and new business transactions. Data thus may constitute a new value. Why Big? The above definition might also be valid for many other types of data and information. Big Data is what you can do only on a large, but not on a small scale in order to gain new insights or to create new values. The preparation and analysis of large amounts of data is done by sophisticated algorithms, applied on these large data volumes in massive parallel processing, run on multiple processors with quick access to these data that are held in the main memory ("in-memory computing"). Thus analysis in seconds—even graphically well supported with "visual analytics"—is enabled. The experts in their respective field "enter into a dialogue" with the data to support their analytical activities. They do not have to optimize questions and wait for the next day until the computer has worked through everything, only to find out that they might actually have asked the question differently. It is easy to imagine that in such a dialogue new opportunities of knowledge discovery arise, which even support intuition. This now seems almost feasible in a way, that is very similar to the functioning of the brain and its intuitive skills using the subconscious (see in addition also the Sect. 5.2.12 for Artificial Intelligence).

In a similar form incoming data streams ("streaming") from multiple sources can be processed continuously and can be automatically analysed in a way that queries and questions can be responded in "real time" by the user, the operator of the respective system or a manager when warnings or irregularities occur. Thus the "subconscious mind of that organization" (or its System 1) evokes the system 2 (following Kahneman) by consulting the responsible human—"the human in the loop", as that is sometimes called in recent literature. In this context we increasingly speak of "Smart Data". This is also an element of a so-called "real time enterprise" (see Sect. 3.2.14).

A particular application of Big Data is the predictive analysis, where from a large number of identified, comparable and similar cases in the past, the likelihood of further development, for example, of disease conditions, is statistically predicted.

Big Data offers three fundamental changes compared with the established paradigms of data and information processing:

- Sample size: N = everything
- Fuzziness in large data collections is permitted
- Instead of the thousand year old search for causality, in which the data has been selected or already collected for a purpose and according to an hypothesis, correlation appears on stage in a new role: correlation partly replaces causality. Big Data enables "Data to speak" and "tell us stories".

The "social graph" of Facebook, which helps to place advertising based specifically on the assumed interests of potential costumers, is another example for Big or Smart Data.

In dialogue with large amounts of data new relationships are recognized and hypotheses are formulated supported by statistical methods with powerful new technologies such as in-memory computing. This helps in the verification or falsification of scientific hypotheses with the traditional research methodology. These now available techniques,—to enter in a dialogue with the data,—are very powerful and will thus drive change in the way how research is done in certain areas. A term that has arisen in this context, is "Knowledge discovery". Another synonym, for example, when large quantities of data e.g., sensor data and other—maybe partly seemingly unrelated—data from various sources are investigated and lead to new statistical context, is "reality mining" as a modification of the already older term "data mining". Davenport (2014) describes in his book "Big Data @work" some application examples in various industries.

David Weinberger analyzes the US Big Data euphoria starting 2010 and goes one step further: In his book "Too Big to know" he defines the need for a new infrastructure for knowledge and the need to have better metadata as better starting points to better link knowledge and to create knowledge (see Weinberger 2011). This brings us to the topic of knowledge management.

Big Data will also change the way the decisions are taken in future: Most institutions are based on the assumption that human decisions are taken based on more or less exact and causal information—whereas in the future—due to the amount of data—decisions will often not be taken by humans but will be taken by machine (see Mayer-Schönberger and Cukier 2013).

This future should be implemented only gradually, in my view: initially we should begin with decision support systems (DSS) so that the human and the human intuition can act as corrective and the systems can still learn from it before decision automata are fully automated in suitable areas. These systems will evolve to parts of the subconscious mind of the respective organization. Their implementation in the organization and the interfacing to the "awareness of the organization" are to be made well-considered and should not just "happen". In this case, the risk of mistakes would be high. More can be found in the chapter on the design and shaping of the subconscious mind (Chap. 5).

3.2.9 Simulation

As we saw, a new way to explore the future is the predictive analysis as a result of the new Big Data technologies. A more traditional way to explore the future is the field of modeling and simulation. Both—predictive analysis and simulation—combined with good visualisation techniques are bound to change the cognitive processes of humans as well as of organizations by giving a picture of possible futures thus influencing decision making. The most common methods are characterized briefly below (see Niessner and Rachinger 2014):

DES (Discrete Event Simulation):

In this method the status changes only at certain times. The considered entities move as actors through a predetermined process. Each entity can be represented with variables. Thus a variety of possible effects and connections can be represented at the level of an individual entity as well as at the level of the total population. This method is suitable for network-based modeling for small populations. By simulation of individuals, results can easily be visualized and understood (for example, in the form of animations). DES is a standard for production, process and logistics simulations.

ABS (Agent Based Simulation):

An agent can be: A computer program, a robot, a human etc. with the ability to perform autonomous actions. The agent is therefore an active part of the model. The agent modifies itself, acts on its environment (he is persistent), obtains information from it, acts in relation to it, has limited perception and action radius (locality) and has a non-trivial behavioral repertoire. The focus of the simulation is on the individual agent and its behavior and control (for example autonomous transport vehicles, increasingly whole supply chains) or the behavior of the mass (in social science issues). Special applications are for example general traffic simulations and traffic simulations at major events.

System Dynamics:

System Dynamics is used for holistic analysis and simulation of complex and dynamic systems. Only a starting value is defined for all parameters considered. The central construct are feedback loops (closed loop). A simulation comes into play when a decision is made under the influence of the information of a certain system status, or when an action triggers the simulation, or if the system status changes. The term "decision" in this context is defined very general with different qualities: aware, unconscious, automated decision (also in biological processes). Application of System Dynamics effects relationships which are difficult to quantify, complex and non-linear dependencies including social, economic, biological and ecological systems, the spread of infectious diseases, simulations for strategic planning, adoption analysis, deduction of consequences in growth strategies of startups, innovation effects, market developments, etc.

Combination of different simulation techniques (hybrid simulation approaches):

Often these simulation techniques are combined, for example, event simulation (DES) combined to simulate a partially automated manufacturing system with agent-based simulation (ABS) for autonomously acting transport units such as cranes, autonomous vehicles (AGVs) etc.

Models and simulations thus are excellent for "Team Learning" and for communication of complex issues. Herb Simon, the first Nobel Laureate in Economics, has said in 1980 that he considers the application of computer-based mathematical models on complex social problems one of the greatest invention of all time (following Carnegie Mellon University 2001). The model is the medium of communication to share the aggregated knowledge in its full depth with the group or the organization. All assumptions can be communicated to top management, sponsors, boards of directors, investors, etc. as a coherent model and in a harmonized way of looking at the key figures. In order not to induce too great expectations, Herbert Simon should be quoted again: "All models are wrong, but some are useful".

These classic simulation methods are based on causality-based relationships that are represented in the model and thus are causally comprehensible. Model types are e.g.,

- Predictive models like weather simulations
- Epic, narrative models like scenarios for alternatives of the future, sensitivity analysis etc.

Models and simulations sharpen our intuition as compass and navigation instruments. Models as stories and metaphors of the real world help us to explain sequences of events and their consequences, help us to focus our attention at a situation, let us learn faster etc. Strategic decision support requires a repeatable process and a model-centric process (see Ghasemipour-Yazdi 2014).

As a basis for simulations in the economic sector serves a cross-industry standard for data mining (see CRISP_DM 2013). The continuous improvement of the business model is in the foreground. Improving decision-making is done by the models and simulations. They are thus used as a "wind tunnel for innovations and decisions".

There are opinions that the correlation analysis with Big Data technology approaches would replace causality oriented simulation techniques. I think that a good combination of the two disciplines and a good selection of the right method are crucial in order to achieve high quality simulations and decisions. In the area of Big Data technologies and their "toolbox" of methods and algorithms we can find for example functions for model generation in the form of decision trees. Decision trees develop from the statistical analysis of raw data and self-learning features which improve the decision tree with numerous "runs", until the improvement from one "run" to the next is only marginal. Then the decision tree model can be used as a prediction tool for new cases. By analyzing the obtained decision tree, experts can possibly also recognize new causalities (knowledge discovery) and teams can improve their knowledge and their sense of reality for future decision making. It is obvious that these new tools are on the way to change how decisions are taken in future be it deliberate, semiautomatic or automatic—they are changing the subconscious mind of an organization.

3.2.10 Privacy and Data Security

Our society has long experience in the assessment and regulation of human behavior. Our today's laws originate from that. But how to regulate an algorithm? How to ensure data protection in the context that it is a feature of Internet culture—especially with young people—to provide personal information voluntarily in the net. Unauthorized access has to be consistently punished in order to enforce the awareness that data privacy violation is a criminal act. In some European countries, this is indeed already working. The possibilities for the detection of unauthorized access are already manifold especially in some areas such as healthcare, where logged access is in many cases already standard. The ability for pattern recognition, which is increasingly available with Big Data technologies will make it easier to identify suspicious traffic patterns that indicate privacy violation.

The risk is shifting from privacy violation towards the assessment of individuals on the basis of probabilities based on the algorithms of "predictive analysis"—something out of the control of the potential victims. Philosophically, the question arises as to the role of free will against the dictatorship of the data. The freedom of the individual to choose and to decide seems to be in contrast to the forecast, based on "Big Data". Just think of the movie Minority Report with Tom Cruise, in which—due to the crimes predicted—individuals and persons were focused on and arrested who had not yet committed a crime. If such scenarios become feasible—and in the next few years much in the field of predictive analysis is going to be developed,—then we will probably also need new rules to protect individual freedom.

In medicine models are conceivable in the form of a "virtual patient" ("digital twin") in their "appearance" (laser scanning and motion analysis), their "internal phenotype" (by medical imaging methods) and their "biomedical phenotype" (pathology, histology, cytology, biomarkers, genome, epigenetics and microbiome). They all will be represented in terms of data. Here too still is some room for discussion, from the perspective of data security and data protection.

The risk of manipulations of the subconscious mind of organizations is of course higher, the higher the degree of "automation of decisions" is. When designing the subconscious mind, it will be crucial to also consider these issues of data privacy and security.

3.2.11   E-Learning, Gamification

Children and young people learn when playing. This also influences their subconscious and develops their skills. The gaming industry has always been a trendsetter concerning user experience. Dealing with complex systems requires an appealing and simple user interface, coupled with the motivation to work with the system in a positive mood. More or less subtle reward mechanisms are one way to increase user involvement and user engagement not only in work environments, but also in learning environments. Models, simulations with different scenarios (starting and input parameters) and predictions from Big Data blended with elearning and gamification have a high potential for shaping the subconscious mind of our organizations.

The consulting firm Harris (2014) predicted that in 2015 50% of the organizations, who have to manage innovation processes, want to gamify these processes. The market will grow from $ 2.6 billion by 2016 to $ 5.5 billion by 2018. A good game design will appeal to emotions and instincts, but naturally has a different impact for different generations—digital natives or digital immigrants. Personal financial management applications, marketing-oriented applications, such as online financial services and online banking portals are possible areas of application for it.

E-learning with gamification has the potential to increase the effectiveness of training, and even to convey fun when learning. There are specifically built-in reward systems in this game mechanisms. Thus awarded points are compared with benchmarks etc., which in turn require an intrinsic and extrinsic motivation of the learners. Such learning modules can be used on mobile devices, when the units are dismantled and offered in small "portions".

These mechanisms are also incorporated in "micro-learning" settings. Microlearning is said to be a very intense and effective learning strategy (Lindner 2008).

Since the motivational systems and structures are a part of the subconscious mind of organizations, the targeted use of these tools is a design option for the subconscious mind of organizations.

3.2.12   Expert Systems, Decision Support Systems, Knowledge Management

Every day decisions on many complex issues are to be taken in organizations at various levels and in various tasks. For decades, the industry was committed to support at least the simple decisions or even to automate them. Thus expert systems for various purposes have been developed, most of which were constructed based on rules—"rule engines"—and they could handle only subjects of limited complexity. Besides this was very costly. The rules try to represent causal relationships involving certain simplifying models of reality. They require high effort for their formulation, development and implementation in the form of decision support systems.

Guidelines, decision rules, decision trees etc. were implemented by "knowledge engineers". In some areas, such as in medicine, these expert systems had very limited success. In practical implementation the complexity of the human organism and its interdependencies to social and physical environment turned out to be too overwhelming as to represent this complexity in models in a sufficient quality.

Big Data now opens up new additional possibilities, namely the building of models from large amounts of historical data with statistical algorithms in order to predict anticipated developments and then, possibly to submit rule-based decision proposals. In many cases—given the predicted most likely future development—the skilled person will already generate ideas while interpreting this projected development—basically understanding the algorithmic reasoning behind. In other cases enhanced countermeasures could be taken in order to avoid the predicted undesirable development or at least to mitigate the impact in a better way.

A very vivid and—positively and negatively—very disturbing example of decision support with "predictive analysis" is "predictive policing". This involves e.g., to support the decision, in which regions and streets of a city the police patrols should be made in what frequency. So targeted policies were enforced on certain streets, on groups of people and on individuals in the USA in Memphis, Tennessee —based on forecasts from past data related to incidents, social data etc. The crime rates could be successfully reduced building on this form of targeted policing (For prediction based policing see Vlahos 2012).

Alternatively, consider the numerous legistic rules and administrative regulations for preventing harm for health from dangerous behavior: seat belts, smoking bans, screening etc. For example it would be technologically possible for insurance groups to extract information from the costumer's behavior as consumer—as far as it is displayed on the Internet—in order to predict the customer's risk for certain diseases (This "social graph" in combination with the information about the health of an individual customer could serve as a basis for targeted advertising). Thus very individual insurance premiums could be suggested by the system. The danger of discrimination against individuals and groups is very high, if not corresponding legal regulations become effective.

Expert systems and decision support systems are part of the knowledge management of companies, which is often automated and—with its still further driven automation—might develop an individual life in itself. Its effects and developments would not be directly visible and therefore would no longer be controllable. These new Big Data-based technological capabilities might strengthen the blind trust in data.

But data sometimes do not record what they pretend to quantify. For example, Robert McNamara was already known as head of Ford for being a very data-driven rational person and he was successful with it. Later he was engaged as Minister of Defense in the Vietnam War. In the assessment of the situation and the actions taken, he took strategic and tactical decisions based on data reported by the individual combat units about the numbers of killed Viet Cong. The fact that the units wanted to be successful for their motivation and considering the fact that a count in combat action is difficult anyway, lead to high numbers of killed Viet Cong reported and thus produced a distorted picture of the situation (see McNamara and Vandemark 1995). That can be considered as a bias coming from the subconscious mind of an organization—the US Army.

There is always the danger that we rely on the results of our analyses blindly—but the data sometimes do not record what they pretend to quantify. "Garbage in—garbage out" also applies with Big Data. But it may have severe consequences, if there are not statistical uncertainties in both directions, which neutralize each other —especially in large amounts of data. Therefore, it is important to identify a potential bias in the data foundation. This requires a good evaluation and assessment of the facts as well as the experience, the knowledge and the access to knowledge. Many organizations are struggling to bring under one roof their content creation, their collaboration within the organization and with their environment and their central business applications and business processes—facing high system complexity and device diversity (Weßels 2014; Weinberger 2011). That is why—in the design of the subconscious mind of organizations—knowledge management is an essential management task and organizational challenge for the future.

Figure 3.1 puts the decisions and the actions to be taken in the focus of knowledge management, embedded in corporate and organizational culture, embedded in methodologies how to innovate, design, evaluate and how to take decisions. The tools for knowledge access and the accessible knowledge base (outer rings) with their interconnectedness to the environment of the organization complete the picture of what knowledge management has to develop, manage and to continuously adapt and improve.

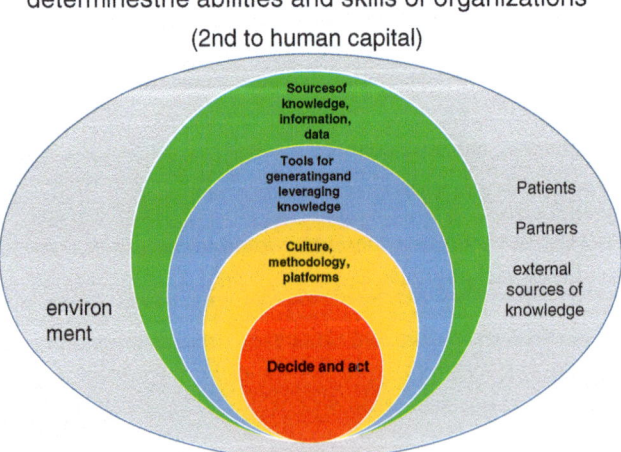

Fig. 3.1 A model of knowledge management

That shows, that it is in particular the use of decision support systems (DSS) in an organization, which is essential to the subconscious mind of organizations. These decision proposals may be given the nod uncritically, they can be implemented automatically, they can be scrutinized and be constantly improved, etc. To design and shape this in the best possible way—following the mission of the organization—will be increasingly critical to the future development and success of the respective organization.

3.2.13 Automation, Robotics

Starting point of the third industrial revolution—automation—were the first programmable logic controllers and thus the advent of electronics and information technology in the early 1970s. Before was the second industrial revolution in the early 20th century with the introduction of mass production using electrical energy. This 2nd industrial revolution followed the first industrial revolution, which had begun with the introduction of mechanical production using steam and hydro-electric power—starting with the invention of the mechanical loom in 1784.

Now we speak about "Industry 4.0"—the fourth industrial revolution—and sometimes in the Anglo-American region—about "Automation 2.0".

The basis for this are "Cyber-physical Systems" (CPS). CPS has become a common term, that describes the connection of sensors and actuators in production with the control and planning systems on the one hand and the products identified electronically on the other hand, bearing the "work plan" in themselves and knowing what processing steps they need next, and in what steps they will be incorporated into the resulting product. This is done partly using the Internet of Things (IoT) and it additionally requires the communication and the right algorithms for self-control of such complex systems. The standards in order to allow such a "cooperation of things", are just emerging and will speed up the implementation of these options.

Any resultant comprehensive data from operation, condition and environment data can be used to produce more efficiency. Further the products themselves can be enhanced for example machine tools, tool holders, etc. can be equipped with Industries-4.0 technologies and thus can provide new service models concerning maintenance or pay per performance remuneration models can arise. With industry 4.0 the complex production processes can be monitored in near real time. This creates greater transparency and reduces storage costs. In addition the globally distributed production processes can be easily replicated, adapted and customized according to the idea "plug and produce". Additive manufacturing and 3D-Printing have the potential to decentralize production in the "garage", at "maker bots" in smart workshops and future labs or "fablabs"—at least when only small lot sizes are required. On the other hand entirely new structures can be printed from food to spare parts and even the first houses already have been printed.

Production processes and products can thus be replicated easily and the production can be adapted quickly and flexibly to the respective requirements. Products can be manufactured more customized and on demand than previously.

Robotics will play a special role in production—as well as in services—as a combination of the advanced capabilities of the sensor technology and mechanics with the information and communication technologies. Formerly industrial robots for quite complex but recurring automation tasks such as welding robots and assembly robots in production were the vanguard of the technological development itself. Today it is the self-guided robot like AGV's (automatically guided vehicles) in warehouses, minesweeper robots and drones in the military and many different types and kinds of industrial robots. Telepresence robots, telemedicine robots, surgical robots and even care robots—particularly in the demographically aging countries such as Japan—are obvious to a breakthrough. Special care robots like "paro" are specialized on affective computing, realizing moods from speech recognition and speech analysis as well as from pattern recognition of what Paro perceives as pictures, movements or contacts from old people carressing their care robot.

This is not only going to change organizations and markets concerning production and delivery processes, but this requires new collaboration and decision processes not only within an organization but also in the cross-company collaboration processes—in the value chain. So the subconscious mind of an affected organization will change at least as reaction to these developments. It is obvious that it would be wiser to anticipate this and shape and form the subconscious mind of your organization by anticipating and considering these changes profoundly. The following chapters might help you with that.

In science fiction movies, the robots are already produced, serviced and programmed or controlled by robots. Until then, it will take some time, but the way to get there and how this development is carried out on a human and social level remains to be seen. Will these major developments be controlled democratically and politically aware?

I think concerns in this respect are quite understandable.

Closely related to the field of Robotics in development, programming and control are the areas of virtual reality and augmented reality as described above. In connection with the advanced sensor technology they enable humans to expand their tools and they even enable manual actions taken over great distances. Drivers in these developments were the (computer) game industry and the film industry with their animated films and science fiction films.

In the combination of these developments we can increasingly speak about "Smart Services".

The Oxford economists Carl Frey and Michael Osborne in a study of 2013 emphasized the fact that—while the brain of the robots have only obediently processed what the engineers dictated them—they are now more and more able to learn from the past and to decide (see Frey and Osborne 2013)—with all the impact on the world of human labor and society. Google's cars are already driving automatically in the United States. Intelligent software checks the creditworthiness of borrowers, screens mountains of documents and files, etc.

The two MIT economists Erik Brynjolfsson and Andrew McAfee—in their book "Race against the machine" (2011)—explain why for the first time since the invention of the wheel the progress destroys more jobs than it creates. The promise that innovation brings greater prosperity and higher wages for the workers, no longer applies. Faster than the advance in computer processors is the progress of the software algorithms. Our education systems, vocational training and adult education have to teach skills how to benefit from the technology instead of leaving people exposed to being threatened by technology. "It is a race technology against education" (Brynjolfsson and McAfee 2011).

If one continues to think of these scenarios, you end up with Ray Kurzweil, one of the pioneers of Artificial Intelligence, who has derived in his book "The Singularity is near" (2005) something like the takeover by Artificial Intelligence. That is far ahead, so let us try to focus on our respective organization and consider what this "here-now-for me"-attitude of costumers and employees, the personalisation enabled by social media, the "sharing economy" and this "instant economy" including the new possibilities of additive manufacturing (3D-printing) means for my respective organization—does my organization have to develop towards a "Real Time Enterprise"? In case of that—what will my organization have to perform in its subconscious mind? What is left and what is necessary to perform more consciously?

3.2.14 The Real Time Enterprise

Damasio (2012) claims that in the course of evolution the processes of consciousness have become more complex due to the benefits in evolutionary competition that developed from functions such as memory, reasoning and language. This made it possible to overlook the near future, to delay or to prevent automated reactions and responses—proposed by the subconscious. This allowed not only immediate but also delayed reward and punishment, and it allowed the estimation of future benefits and disadvantages.

It is from this perspective crucial, that organizations and especially companies being in competion have a clear view of reality in order to act as immediately and intelligently as possible.

Today a digital competitor can quickly appear on stage. The advances in cloud computing have made the necessary technology faster available for all market participants—both at the customer and consumer front as well as in the back office. It is no longer necessary for new competitors to make massive investments in their own systems. The technology and resources can quickly be utilized from the net. Think for example of 3D-printing.

When you want to be successful as a realtime enterprise in future simplicity, flexibility and agility have to be top priority. Visualization of reality in a direct dialogue with large amounts of data instead of "monthly analysis based on complicated aggregated data" is now possible with the new technologies.

Also, the direct involvement and processing of incoming data ("streaming data") is possible, e.g., when just being involved in an online purchasing process with a customer, having access to the customer data and to available data of similar costumers, an offer or additional similar products (possibly already available) can be displayed in order to complete and fix the order as fast as possible and specifically tailored to the costumers demand. The real-time enterprise in real-time dialogue with customers.

Many such scenarios and operations can only run automatically in the background—in the "subconscious mind of the organization", so to speak. Specific developments and abnormalities have to be recognized immediately and the system 2 must be evoked being capable of making decisions, consciously involving the people in charge—thus improving the quality of the real-time enterprise continuously. Or—in the case of a massive aberration or threat—you can prevent harm to the organization. As you see the design of the "subconscious mind" of such organizations is of particular importance on the way through the digital transformation.

3.2.15 Artificial Intelligence

I want to focus on Artificial Intelligence (AI) now, which represents a spearhead of technological possibilities and which will be of particular importance for our organizations as well as our future civilization. That is why you will find some far-reaching and philosophical considerations in this section—from various sources: poets, technologists, entrepreneurs. I want to put a very specific as well as fundamental focus on this core topic of this book: the interrelation of the consciousness and awareness of organizations to their specific "subconscious mind" with Artificial Intelligence as a key technology incorporating many of the technologies described above. This topic is also key to the second core theme of the book, the collaboration of man and machine forming "hybrid intelligencies" woven in the fabric of future organizations (see Chap. 6).

In a long tradition of various sciences and schools of thought, such as psychology, cognitive science, neuroscience, religion and philosophy, mankind has tried to fathom the human mind and to work out what characteristics distinguish it from other forms of life. An essential topic is the role of the brain and the nervous system—often discussed under the concept of "mind-body problem". This raises the question of which living things do have something like a soul or a spirit at all. In English literature the comprehensive term "mind" describes several concepts in an overlapping way that have distinctive meanings themselves like soul, spirit, mind, consciousness, intelligence.

On the one hand the demarcation between the human intelligence from the animal world is discussed. On the other hand that also raises the question—and this is interesting for this book and the Artificial Intelligence in general—of whether not living matter, like for example man-made machines can be awarded such characteristics like a "mind". A spirit, a soul enables humans to feel subjective perception and to act in pursuit of a will as well as to put himself in relation to and to perceive his environment ("intentionality"). It enables humans to react in a certain intention on stimuli, as well as to have consciousness, thinking and feeling (Oxford American College Dictionary 2014 on "mind").

In the late 20th and resulting in the 21st century cognitive science has developed several different approaches to the explanation of the mind. The possibility of non-human intelligence is developed and researched, which in turn is closely linked to the information theory and cybernetics in the field of Artificial Intelligence. This is an attempt to understand how human mental capabilities can be emulated by nonbiological machinery.

With animals the brain is the control system of the nervous system. It is close to the primary sensory organs for sight, hearing, taste, smell and balance. Primitive organisms such as sponges do not have a brain. A brain is enormously complex. The human brain has 86 billion neurons, each neuron is connected on average to 10,000 other neurons (see Whishaw and Kolb 2010). Many features of human intelligence as empathy, sadness, rituals, the use of symbols, the use of tools, etc. are available in a simpler form even with primates.

For the development of human intelligence, there are many theories. The emergence of language as a crucial development step is undisputed. The "Group selection Theory" says that characteristics have prevailed in the evolution, where the benefits have been favorable for a group, even though they have brought disadvantages for the individual. Another theory says that intelligence is linked to diet and therefore to opportunities provided by social and physical conditions (Bergstrom 2002).

This Group-selection theory can be applied to organizations in a thought experiment in analogy and would suggest that the new technologies and especially Artificial Intelligence offer new possibilities of integration strength and strong bonding within the organization, but also in the interaction with others. Organizations offer even within their organization a special proximity concerning social and physical conditions.

Now—after these basic and nearly philosophical considerations—let us have a look at some examples of Artificial Intelligence systems:

- Autonomous Vehicles from Google are approved for transport in the US and constitute a system of Artificial Intelligence. Autonomous vehicles that communicate with each other and thus regulate the flow of traffic without any interaction by the driver himself are already in development. What is interesting about this situation is that it may come to constellations where no parent controller entity for all vehicles is involved, but that it is something like "swarm intelligence" where each object may have a different destination—as opposed to the in the nature occurring phenomenon of swarm intelligence as we for example can see with flocks of birds flying in formation.

- IBM has done pioneering work with the system "Watson". Watson won the popular quiz show Jeopardy, gives advice for physicians in the field of oncology and "draws knowledge" from reading numerous medical journals. A precursor of Watson was the chess computer BigBlue from IBM who has defeated the world chess champion in the late 90s and had implemented the most advanced strategies and algorithms for decision-making in view of countless choices. In Watson numerous technologies of Artificial Intelligence in particular the technology of NLP (Natural Language Processing) are implemented—as a prerequisite to support the users in their decisions with context-sensitive Artificial Intelligence. The speech recognition as such was always a driver and a key discipline in the development of systems of Artificial Intelligence. The speech recognition is crucial especially in the field of medicine, where natural language descriptions of medical history, diagnoses, conclusions and therapies as well as dealing with uncertainties are common and where it is not possible to put everything in mathematical models as in other—purely technological—areas such as for example the intelligent control of energy networks.
- The importance of speech recognition in Artificial Intelligence and their potential becomes evident, when we remember the crucial role the introduction of the language played in the evolution of mankind.
- Affective Artificial Intelligence deals with perceiving and processing emotions from humans as well as interact with humans politely and with empathy, given that this is not possible for "deep empathy" etc. In Japan these sort of robots seem already to be quite popular.

Many people wonder: Is our voice-enabled equipment, such as "Siri", the assistant in the iPhone able to think? Are they independent beings? Do we enter a new era of intelligent machines with them?

Alan Turing in 1950 formulated the Turing test as a benchmark to assess whether a system can be considered as Artificial Intelligence or not: A computer and a human interact under observation by a third party in the role of observer. The observer determines by questions, who is the human and who is the computer. If the observer is not able to determine reliably who the computer is, the Turing test is passed.

The poet and lyricist Clemens Setz has asked the following question in an article in the feature section of Die ZEIT (see Setz 2014): Do humans need voice-enabled computer at all? His answer: No—because androids can think for themselves. They are independent beings who have their own language and their own culture. A provocative thesis of a poet (born 1982) and not of a technician or scientist— keep in mind: it comes from the generation of digital natives.

The American writer Philip K. Dick parodied the Turing Test in his famous novel "Do Androids Dream of Electric Sheep?" (The novel was produced as a film —"Blade Runner"). The book is about replicants that can be convicted as non human by the so-called Voigt-Kampff test. This test does not measure the intelligence, but the compassion. That brought an American research team in 2005 to the idea to recreate Philip K. Dick as a robot. The robot should be not only a remarkably genuine copy of the appearance of Philip K. Dick, but a voice-enabled, intelligent android, which behaves as if it were the late Philip K. Dick. So they gathered all the utterances of the hugely prolific author, stored them on a server, combined it with software developed by the programmer Andrew Olney, and thus enlivened the head of the android. As "Phil" was finished, Philip K. Dick's daughter Isa conversed with the android. It really felt as if she had spoken to her father, she said.

In December 2013, Michael Scherer, then managing editor of Time magazine, was called by an automated software. The robotic voice called itself Samantha West. They wanted to sell him something. Scherer drew on the conversation. When he asked if she was a robot, she replied: "I am a real person, can you hear me?" The robot tried to pass the Turing test—and failed (see Setz 2014).

Setz now asks philosophically whether we should not stop to try to replicate human intelligence, but to simply allow humanized intelligence. The Lingodroids are a new species of robots that converse with each other. Researchers at the University of Queensland have demonstrated that they are able to develop words and concepts, claims Setz. The latest generation, called iRats develop a language that even humans can learn and the iRats develop their own genuine culture. Setz believes the future of Artificial Intelligence lies in the emancipation from mankind. Voice robots, which—like the iRats—are left to themselves, will make a real alien intelligence, regardless of the absurd demands of the human world of communication. They do not sound like us, they do not conclude like us and they do not err like us. But as long as we hold intelligent beings, even if they consist of inorganic matter, on a short leash of humans, they will not get beyond the level of overbred pets, says Setz.

Let us go back to actual but far reaching research: The research project "Google Brain" has developed a completely new principle to write intelligent software by renouncing the connection to all logical processes of human circumstances and let the program follow the dynamics of neural networks in an unsupervised learning mode. This system has been confronted with unclassified data volumes from the Internet. It built up new connections on its own—abstract concepts. One of those resulting concepts which in its way emerged as pattern represented in neurons which were visualized then, looked like the face of a cat—to the great surprise of the developers. No one had given the program the explicit order to inform itself about cats. It recognized by itself what looked like a cat. The internet is full of pictures of cats. An alien intelligence, taking a deep look at our Internet, could assume that cats must be a foundation of the entire world's structure. That is why the program Google Brain making its own discoveries in an unsupervised mode focused on cats. It proceeded like an archaeologist who explores the mythology of a previously unknown culture searching for patterns, clusters etc. without bias (see Setz 2014).

Clemens Setz suggests conclusively that the digital Adam should be set aside a digital Eva. "No creature deserves to be left alone" he says. It might be fun to think about your organization in these "artificial" categories—maybe you will end up a bit confused in thinking about the future of mankind and you will have lost the structure of organizations as well as other societal structures as we know them today.

I consciously put the idea of a writer and poet about "Artificial Intelligence" first and now will add the view of renowned technologists on this topic: Ray Kurzweil with his "pattern recognition theory of mind—PRTM" with the concept of neuronal networks being the core of his book "How to create a mind" (Kurzweil 2012). Based on this theory he tries to build an understanding about the brain and its functioning in a re-engineerig process and to consider how such a thing can be designed. I try below—based on Kurzweil's idea and presentation to give an impression how we can imagine that and what strategies Kurzweil chooses (see Fig. 3.2).

The central element is a pattern recognizer (PR)—a pattern recognition module —built as a neural network. These PR modules are interconnected with adaptive genetic algorithms, each axon is connected with one or more PR modules of the next higher level thus forming virtual circuits—which is more flexible than in the human brain, which indeed physically forms these connections.

The system is enabled to create as many PR modules—in as many levels—as necessary. This system forms out its own topologies, depending on which external patterns it is exposed to. It also allows redundancies in order to enable a pattern recognition as robust as possible. The learning process is a gradual one—created in massively parallelized form step by step and from one level to the next level.

The purpose of the artificial brain is defined with an input in the form of targets —such as to pass the Turing test for example. A "review module" performs background scans of all existing patterns in order to verify its compatibility with

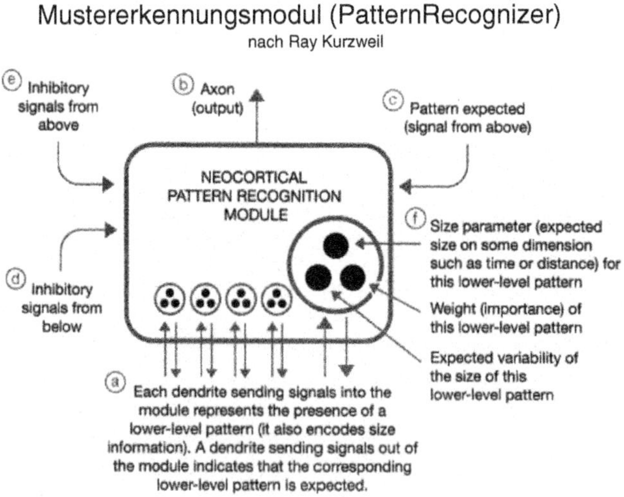

Fig. 3.2 Pattern recognizer. Ray Kurzweil, How to create a mind, Viking Penguin group

other patterns. Another module determines questions and feeds it to other knowledge areas of the artificial brain.

Kurzweil defines intelligence as the ability to solve problems with limited resources. Intelligence has evolved because it has ensured the survival in evolution. The development of our human species due to our intelligence is reflected in the evolution of our knowledge, our technology and our culture. The technologies are increasingly of information-technological nature and undergo an exponential development, particularly due to the increasingly rapid replicability in terms of algorithms and software—and to the "cloud" in the background. Humans "extend" themselves virtually in this way. Much of our knowledge is already in the cloud (see Kurzweil 2012).

Kurzweil predicts that we grow together with our technologies—we merge with them. With the cloud of knowledge we have already done this virtually. Regarding our perception this development has already begun with virtual and augmented reality (VR and AR, see above). Physiological compounds with implanted chips will follow.

If we—with our human intelligence—succeed to gain access to this non-biological intelligence and "extend" ourselves this way, further exponential developments will occur—according to the LOAR ("Law of accelerated returns"), postulated by Kurzweil in the information technology area and derived from recent developments. The brain is gaining access to its "source code" by this reverse engineering and will have the opportunity to improve—accelerated in iterative cycles.

With this expansion of our intelligence to Artificial Intelligence, we will be able to achieve ever higher levels of abstraction. An ultra intelligent machine could still produce better machines—there will be an explosion of intelligence—the singularity? One of the main works of Ray Kurzweil is "The singularity is near" (Kurzweil 2005).

Some of these scenarios will probably also remind you of science fiction movies. Of course there are many critics of these hypotheses and scenarios from the field of Artificial Intelligence, such as Paul Allen and Mark Greaves. They argue for example, that the LOAR will only work until it no longer works. On the other hand they contradict the hypothesis about ever accelerating software improvement, arguing that software developed slower than hardware in the past, etc.

Today we deal with specific AI (sometimes named "narrow AI") with each system trained and focused on a specific area for specific tasks like decision support for medical diagnoses for cancer, playing games like chess or Go, move robots in specific surroundings with specific tasks etc. Narrow AI clearly beats the human in many specific tasks. We use them as tools—hoping they won't get too autonomous and replicate themselves in an uncontrolled way. When we speak about emulating the human brain or even approaching the "singularity" we speak about "general AI" which indeed will be a long way. Google and others already seem to be heavily investing to be the first and then presumably dominant player in this field.

Of course thus we arrive fast at the border region to issues of politics, philosophy and religion. Where does this journey lead us? And how fast does it go? What will be the development of the human mind? Who will have access to these resources? Who will influence the development? What's about our consciousness? What's about the free will of the individual, will it still be there? What will happen with the identity of the individual? These questions are not subject of this book. But it is an issue that we must somehow position ourselves in our organizations and organize so that we, organizationally and ethically deal with these technologies and their potential for Artificial Intelligence in full awareness what we want to allow in the background—in the subconscious mind of the organization. As many features of "Artificial Intelligence" will "creep in" there, we will be challenged to recognize this and to learn how to properly use it, in terms of the design and the shaping of the subconscious mind of the organization (see later in this book).

Peter Thiel, founder of PayPal payment service and major investor is,—as far as the general social effects are concerned—not so pessimistic about the dangers of Artificial Intelligence as some other authors. Concerning the coexistence between humans and computers, and the threat to jobs—often to those in the West—derived from the experience with globalization and the competition between workers in the west to those in India and China, he says "technology" is different. Computers and humans are very different. Computers can make huge calculations, humans are good at making judgments and decisions, at making plans, in understanding the meaning of things. Humans and computers are no substitute for each other to a large extent. Technological progress creates more opportunities for the people than it destroys (Heuser 2014).

After this "visit" to philosophical views and to possible scenarios, I would like to take a look—as an example of "narrow AI"—at one practical application of a basic technology of Artificial Intelligence—to neural networks. This example comes from the area of power supply (see Siemens 2014):

Wind and solar energy have got the energy markets moving: Now no longer does only electric energy consumption fluctuate but also the amount of electric energy available. How to deal with this new situation? The answer is: Improved planning through good prognoses. If you know how much solar and wind energy is ready the next day, and you have forecasts of the regional demand, you can control conventional power plants with foresight, planning enough balancing energy for losses due to electricity transmission and securing the required quotas at favorable terms on the power exchange.

Therefore forecasting and simulation software based on neural networks is required. Neural networks are computer models that recognize relationships through training based on data from the past and thus make predictions. It is not required to fully and analytically understand the problem. For example, weather forecasts for previous periods and the electricity production of a solar park in this period serve as a basis to simulate future production from the current weather forecast. Initially, the program does not know how the parameter affects the electricity output and if the forecast deviates significantly from the produced electricity. During training, the program minimizes the difference between its forecast and the actual value in thousands of runs. It changes the weighting of the individual parameters and the neural network becomes more and more accurate.

This technology has been around for 20 years. With the recent technological developments in the hardware and software area (see above, Big Data in Sect. 3.2.8, simulations in Sect. 3.2.9, etc.), these neural networks now apply to more complex issues and are much faster and more powerful than earlier. When we then take the electricity generation patterns of the various systems including the characteristics of the individual solar parks and combine them with the electricity consumption patterns of major regions to be supplied, you need these advanced capabilities of powerful neural networks to balance the net. So you can, for example, also schedule maintenance work of power generation facilities (for example, cleaning dirty solar panels) in times when their energy supply will be least missing.

3.2.16   Megatrends, Predicted and Propagated by the Consulting Industry

Before going to build a model of the subconscious mind of organizations in Chap. 4 , let us now try to summarize the relevant key megatrends from different perspectives, as they are regularly provided by the major consulting and analyst firms, such as Gartner in the field of information and communication technologies (see Gartner 2013). Following, you find a list of relevant Megatrends concerning the shaping of the subconscious mind of organizations and development of "hybrid intelligences", supplemented by a brief explanation (In the previous chapters a greater focus was given to some particularly relevant technologies):

- **Increasing variety of devices**:
  Starting with the PCs, on to laptops, tablets, smartphones, wearables, smart tables etc.
- **Mobile applications**:
  Increasingly, applications are running on mobile devices. These so-called Apps are the building blocks of new application architectures and are supposed to secure the interaction between different functions and tasks via standards. This still will require some more effort. Thus, the interaction between objects will become increasingly important and will thus partly form the subconscious mind of our future organizations.
- **Connectivity**, "**connectedness**" enabling new ways of **personalisation, customer integration**, etc.:
  Things, products, information and locations will increasingly be digitally present and interconnected. People already have themselves widely interlinked with the help of mobile applications and social networks. They have done this also off the networks of the organizations in which they work. The networks of things, information and locations (GPS with associated video, Services,

Facilities etc.) will be integrated into the network. This comprehensive connectivity and the ability to interact with anybody and anything will increasingly be used by the "digital natives". They grow up with these technologies as part of their consciousness and their everyday life. They will use this new way of "being connected" in a not predictable form. So these tools will augmented and extend their abilities and they will shape and influence—as other tools in the development of our civilization also did—the subconscious mind of these "digital natives".

New forms of customer integration (costumer engagement) and in particular the patient involvement in healthcare (patient engagement) will change whole industries, offer new business opportunities and allow other old business models to die. That is what the digital transformation is all about. The "connectedness" in the personal sphere (for example, the "Quantify self"—movement with the continuous recording of vitality and performance parameters in daily life and in sports) will on the one hand change the individual's behavior and will on the other hand lead to the increasing individualization of health prevention etc. It will be increasingly difficult to find the right balance between privacy and connectedness and to allow people with their individual wishes to go their way. It will require a comprehensive and improved legal framework for that.

- **New working environments**: "**New Work**":
  Shortages of skilled workforce will be with us in the coming decades from the demographic and educational point of view. The "war for talent" has in many areas already begun. Keywords of the new world of work are: Lifelong learning, knowledge-based society, "always on", creative industries, corporate social responsibility (CSR), new business opportunities, flexibility, E-Mobility, "selfness", single-person-company, real-time enterprise, etc.
- **Artificial Intelligence**—"**smart machines**" (see Sect. 3.2.15):
  The skills of people will be extended by machines. Machines are replacing humans. The machine will get to know the people and their environment better. People will get to know the machines and their possibilities and limitations better. New technologies like 3D printing or additive manufacturing will change structures of the producing industry with more decentralisation. Machines and people will—hopefully—work side by side and work together—maybe establishing hybrid intelligences with the human in the driver's seat (see Chap. 6).

This will result in significant changes in numerous industries, Some of these trends have the potential to induce and lead to veritable upheavals—not only in the IT industry.

A new global digital economy is emerging (see Oxford Economics 2011) which will expose numerous sectors, industries and companies to a massive pressure for change and require for the organizations to adapt and invest in their methods, tools and processes. Organizations have to be designed and built in a way, so that they can handle these uncertainties, upheavals and changes in the best possible way. Even we as humans are highly dependent on our capabilities and the settings of our subconscious mind in our dealing with uncertainties in surprising situations. Our subconscious mind frames and influences our conscious behavior in many ways—as the conclusions from the perspective of behavioral psychology and cognitive science impressively show. Something similar will happen with organizations. Built on this metaphoric view the thought model and framework for this "subconscious mind of organizations" is elaborated in the following chapter.

Literature

Bergstrom TC (2002) Evolution of social behavior: individual and group selection. J Econ Perspect 16(2), American Economic Association

Biswas P, Aydemir GA, Langdon P, Godsill S (2013) Intent recognition using neural networks and Kalman filters. In Human computer interaction and Knowledge discovery in complex, unstructured, Big Data—workshop proceedings Maribor July 2013. Published in Lecture notes in Computer science, Springer, LNCS 7947

Bruck P (2008) Microlearning & capacity-building, (ed. M. Lindner). Innsbruck University Press, Innsbruck, Austria

Brynjolfsson E, McAfee A (2011) Race against the machine. Digital Frontier Press, Lexington, Mass

Carnegie Mellon University (2001) A tribute to Herbert Simon. http://www.cs.cmu.edu/simon/all.html

CRISP_DM (2013) www.crisp-dm.org/ed

Damasio A (2012) Self comes to mind: constructing the conscious brain. Vintage Books Edition

Davenport TH (2014) Big data@work. Harvard Business Review Press

Elsberg M (2012) Blackout—Morgen ist es zu spät. Blanvalet Verlag

Frey CB, Osborne M (2013) The future of employment.http://www.oxfordmartin.ox.ac.uk/downloads/academic/future-of-employment.pdf

Gartner Identifies the Top 10 Strategie Technology, Trends for 2014, Gartner Symposium/ITxpo 2013 October 6–10 in Orlando

Ghasemipour-Yazdi R (2014) Simulation von Geschäftsmodellen. WINGbusiness 1/2014, S 20

Grüter T (2013) Offline!—Das unvermeidliche Ende des Internets und der Untergang der Informationsgesellschaft. Springer Spektrum

Harris T (2014) The power trio integration through SoA, Enterprise mobility and gamification. Bangalore 2014 www.thbs.com

Heuser UJ (2014) Funkelnder Kapitalismus, Die Zeit, 39/2014 vom 3.10.2014, S 26

Hofstadter D, Sander E, Held S (2014) Die Analogie: Das Herz des Denkens (1. Auflage). Tropen-Verlag 2014 Verlag C.H. Beck im Internet ISBN 978 3 608 94619 2

Klausnitzer R (2013) Das Ende des Zufalls—Wie Big Data uns und unser Leben vorhersagbar macht. Ecowin Verlag Salzburg

Kurzweil R (2005) The Singularity is near. Viking

Kurzweil R (2012) How to create a mind—the secret of human thought revealed. Viking Penguin Group

Lego (2014) http://www.lego.com/en-us/mindstorms/?domainredir=mindstorms.lego.com. August 25, 2014

Mayer-Schönberger V, Cukier K (2013) Big Data—Die Revolution, die unser Leben verändern wird. Redline Verlag

McNamara RS, VanDeMark B (1995) In retrospect: the tragedy and lessons of Vietnam. Random House

Medical informatics world conference (2014) Vortrag von Kaiser Permanate. Boston, USA

Niessner H, Rachinger P (2014) Unterschiedliche Simulationstechniken im Praxis-Einsatz. Top-thema: Business Model Simul 6–9 http://www.google.at/url?sa=t&rct=j&q=&esrc=s&source=web&cd=5&cad=rja&uact=8&ved=0ahUKEwjL0qSd797UAhUBchQKHV0DDVoQFghAMAQ&url=http%3A%2F%2Fwww.wing-online.at%2Ffileadmin%2Fuser_upload%2Fwing%2Fwing-business%2FEditorial_Inhalt_Heft_01_2014.pdf&usg=AFQjCNFldIl-LFsRTpO9eRxaqhQ2h-SEbQ http://www.wing-online.at/fileadmin/user_upload/wing/wing-business/Editorial_Inhalt_Heft_01_2014.pdf.

Oxford American College Dictionary (2014) "mind", Putnam Adult

Oxford Economics (2011) The new digital economy—how it will transform business. White paper from a research program sponsored by AT&T, Cisco, Citi, PwC, SAP

Senge P (1990) The fifth discipline, the art and practice of the learning organisation. Doubleday currency

Setz C (2014) Der Digitale Adam. Die Zeit v 10.7.2014

Siemens hitech 2/2014. Wer lernt, gewinnt

Steyrl D, Scherer R, Müller-Putz GR (2013) Random forests for feature selection in Non-invasive Brain-Computer Interfacing. In Human computer interaction and Knowledge discovery in complex, unstructured, Big Data—workshop proceedings Maribor July 2013. Published in Lecture notes in Computer science, Springer, LNCS 7947

Ungericht B, Wiesner M (2011) Individuelle und organisationale Widerstandskraft—Chancen und Missbrauch des Resilienzkonzepts. In Zeitschrift Führung und Organisation, Heft 2/3 2011

Vlahos J (2012) "The Department of Pre-Crime", Scientific American. Bd 306

Weinberger D (2011) Too big to know. Basic Books

Weßels D (Hrsg) (2014) Zukunft der Wissens—und Projektarbeit—neue Organisationsformen in vernetzten Welten, Symposion-Verlag

Whishaw IQ, Kolb B (2010) An introduction to brain and behavior. Worth Publishers, New York

Chapter 4
A Model of an Organization—How Do the Subconscious Mind and the Conscious Mind of an Organization Work?

Abstract As a basis for the design considerations of the subconscious mind of organizations and its interfaces to the conscious mind, a model of an organization is developed—starting from the individual employee, and the individual organizational units, to their interaction in an organization as a whole, supplemented by connecting elements such as working groups, project groups, etc. These elements are embedded and interconnected in the infrastructure of the organization and its internal information and knowledge sources. On the other hand, they are also—increasingly through new technologies—in many ways tied to the spheres of the customers and stakeholders, to the sphere of the market and the competitors and to the sphere of external sources of knowledge and information. The structures are self-similar and cascading in the respective level of the organization. This model of an organization with a focus on its subconscious mind is elaborated and explained with vivid practical scenarios and stories and (in chap. 5) with guidelines in order to provide a metaphoric framework and model of thought for mastering the digital transformation in the respective organization.

Keywords Model of organization · Organizational units · Cascading subconscious minds · Infrastructure · Interconnectedness · Organizational framework · Decision making

This model of an organization serves as basis for our considerations, for our modeling, for our analogies and for our metaphors. It is based on the hypothesis of the book itself, namely to apply the categories of conscious and the subconscious to organizations.

In the following sections, the essential elements of the organizational model are presented with their respective possibilities provided by the expansion of perception, recognition, knowledge, memory, decision-making and action offered by the new technologies on the one hand and the interdependence with the environment of the organization (internally and externally)—facing the new technologies—on the other hand. The behavior within this interdependence and its impact are fully explained when developing these elements of the organizational model. Based on

© Springer International Publishing AG 2017   67
W. Leodolter, *Digital Transformation Shaping the Subconscious Minds of Organizations*, DOI 10.1007/978-3-319-53618-7_4

this, in Chap. 5 the design possibilities are explored especially for the subconscious mind of an organization and its interaction—the interfaces—with the awareness and the "conscious mind" of an organization. Lucid stories and scenarios from a fictitious industrial company, a fictitious retailing company and a fictitious health care organization illustrate this model. In chap. 5 you then find guidelines supported by microstories for shaping the subconscious mind of your organization.

4.1 The Basic Element: The Individual Employee in the Organization

The basic element of an organization undoubtedly is the individual employee (hereinafter individual or person) in the organization with its abilities, skills and experiences aligned on its tasks and activities in the organization. This is of course inseparable from the attitudes, values and other characteristics of this person. The private and professional environment of this person of course very much influences its behavior and in particular the way its system 1 ("thinking fast" by Kahneman, see above) works—and thus also affecting its conscious behavior. Other than the conventional—purely behavioral and psychological—considerations about the interaction of each person with its organization (as elaborated in the scientific discipline of organizational behavior), now the tools and options for enhanced perception, recognition, memory, knowledge access outside the person's brain and new tools of action and communication have to be considered. Of course, the social embeddedness in the network of the organization has also to be considered—the social and professional integration in the context of the organization as well with its customers, stakeholders etc.

Each person has a system 1 (thinking fast) and system 2 (thinking slow) with respect to its activities in the organization. The respective person disposes—both for its system 1 and its system 2—in addition to the resource of its brain also of the technologically augmented scope of perception, recognition, decision and action.

> **Example hospital group—*ICT Infrastructure supports staff members in their patient centered work***
> **Robert is a specialist in a department of internal medicine** at a medium-sized hospital with additional departments for general surgery, obstetrics and gynaecology as well as radiology. In addition to general internal medicine, numerous patients with chronic diseases, esp. diabetes and heart/circulatory disorders, as well as geriatric patients and cancer patients are cared for in cooperation with the Oncology Center of the nearby university hospital. Affiliated to this department are 20 beds for remobilization and aftercare (aftercare clinic) especially for older people who no longer need acute medical care, but are not yet independent and healthy enough to be able to care for themselves at home.

Mrs. Huber, a widow living alone at the age of 78 years had been referred to the Department by her general practitioner (GP) 20 days ago because of an enlarged lymph node. She had moved to the city three years ago after her husband's death and after recovering from breast cancer treatment in order to live near her son's apartment. Her son, who was divorced, had recently changed his employer and had frequent business traveling abroad. Despite these changes she lived independently. She believed that she had beaten cancer after the encouraging results of the checkups and was happy and enjoyed a good quality of life.

Robert had just returned from a three-week vacation and was given an overview of the patients on his ward from his colleague, who had been filling in for him as responsible senior physician for his ward and now had begun her well-deserved vacation. Actually, Mrs. Huber should soon be discharged.

The information technology infrastructure of the hospital had improved massively in recent years. The electronic medical records had already been available for ten years; with elderly patients it was very difficult to scan through the numerous findings and medical reports on various diseases and incidents from the broken finger to cardiac arrhythmia and to obtain the relevant information in the respective treatment situation quickly. Often the physicians tried to evade from this extensive study of reports and filtered out the important and relevant facts in a detailed conversation with the patient. Moreover they could only learn in the medical history of what has happened in their own hospital network, but not whether and which treatments were carried out at other clinics or general practitioners. Since five years there was now a nationwide electronic health record available and thus largely complete information of the patient's medical history. Since the last year there was also a pilot installation available that visualised relevant information for treatment related to the medical context. For example dependent from the situation, if the physician actually was involved for example in emergency room or at a routine check after heart surgery, the physician was supplied with the relevant information for it. Supposedly, the system should be a learning one and constantly improve. The young fellow physicians had gone so far, that they even took treatment suggestions from the system's decision support functions at face value and often—as was the opinion of some of the older colleagues—uncritically accepted these suggestions and acted following them. Deep evaluation and reasoning as well as intuition and "feeling"—developed in a long professional experience—would run the risk of being lost. The older colleagues already began to accept the system as decision support tool and were also encouraged by the management to look at it critically and to actively contribute in order to improve it.

Concerning Mrs. Huber, Robert now used this infrastructure to get a quick overview and saw the essential parameters of the tumor boards, where the colleagues from gynaecology, radiation therapy, radiology, surgery, pathology and colleagues specialized in oncology from his department had discussed the case and recommended surgery with subsequent chemotherapy. As accompanying illness an adult-onset diabetes had been displayed—not unusual for women of this age. The surgical report and the report from the pathologist after surgery were encouraging. Entering the keyword "Social history", Robert saw after seconds that Mrs. Huber lived alone and had given her son in the same city as contact address. That was for Robert already a tremendous asset and an improvement of working conditions as compared to previous years. He could now—without having "digged around" in the electronic medical history for at least 15 min—prepare effectively in five minutes for the upcoming conversation about discharge. Mrs. Huber would be impressed how well prepared the "new doctor" would have come to the discharge visit.

The latest achievement in the ubiquitous information technology infrastructure of the hospital group Robert had learned before his vacation in a training session: Support for the discharge management. The doctors had been trained together with the ward nurses. Based on the millions of case histories of the group—so was the explanation in the training—statistically (in their entire medical history) similar cases were analyzed and the system predicted by statistical methods and algorithms, when Mrs. Huber would be re-hospitalized. After seconds the result was there: the probability was 30% re-emergence of swollen lymph nodes in six months and thus a new hospital need. That Robert would have intuitively expected too. But what surprised him was the high probability of a readmission in three months due to hip fracture. Although it was a purely statistical method—as stated emphatically in the training—the supplied keywords indicated the causality—the possible clinical reasoning—behind: osteoporosis, blood glucose level, risk of falling.

In the case history, after a new request—tagged with "skeleton"—a broken finger was noted but no other fractures. There has been no investigation on osteoporosis.

Thus equipped, Robert went to the discharge conversation with Mrs. Huber. She told him that she had a forearm fracture ten years ago at her former place of residence, that it was well healed, and that she had no problems with the arm any more. So Robert had bone densitometry checked with the result that medical treatment and diet to prevent further deterioration in terms of bone density were necessary.

The **ward nurse Sarah** had also prepared for the final discharge interview with the new system and, in consultation with Robert before her conversation with Mrs. Huber they came to the following conclusions as recommendations for further care:

- Two weeks remobilization and aftercare in the aftercare clinic affiliated to the hospital to ensure the mobility and to lower the risk of falling by special physiotherapy
- Accompanying psychological visits within the aftercare clinic

- Control of diabetes with education in nutrition and behaviourial training specialized for medication adjusted diabetics
- After discharge from the aftercare clinic: Weekly visit by the home nursing unit for four weeks
- After six weeks of care at home: control visit in aftercare clinic focusing on the experience of the patient and control of her handling with diabetes
- Telemedical monitoring of blood glucose level by several measurements daily and telemetric transmission and weekly inspection by GP or telemedical care center
- Drug setting with consideration of diabetes and osteoporosis.

This scenario shows the following relevant characteristics of a subconscious mind of the organization hospital and the hospital group as a whole in a narrower sense as well as the health care industry as "organizational sphere":

- Extended professional "Perception" by information systems in the form of hospital information system and the electronic health record with contextual information processing (*quick and comprehensive context-sensitive support for preparation for discharge visit*) for the doctor.
- Decision support based on predictive analysis in the fundus of the millions of medical case histories (*readmission prognosis at discharge*) as essential extension of "memory" and the knowledge base—in addition to the doctor and their System 1 and System 2 judging the consciously and unconsciously accessible knowledge for intuitive as well as consciously reflected decisions.
- Risk of uncritical adoption of draft decisions by decision support system leading to the loss of "feeling" and intuition (*older doctors, young doctors*).
- The Information Infrastructure supports professional groups and multidisciplinary collaboration because technical hurdles have been removed and collaboration and decision sharing is made easy (*Tumor Board, Robert and Sarah*).

One further scenario:

Example industrial companies—*maintenance of a rolling mill improved by sensors and analytics*
Frank is **responsible** in the rolling mill for **maintenance** of the rolling mill of a steel manufacturer specialized on long products. His goal is to operate the mill without plant malfunction and to keep scheduled maintenance intervals as short as possible. High wearing components such as rolling stands are often available for repairs due to the fact that, when another rolling stand is in operation for rolling its specific dimensions, the other rolling stand is not necessary for production at this time. This is being coordinated by production planning and control. Frank is often himself hands on involved when the rolling stands are disassembled. For more extensive repairs, he has access to

skilled workers from the maintenance organization. Previously they had made mechanical apprenticeships as machinist, toolmaker, steel fitter etc. For the past 15 years, workers trained in mechatronics and electronics also appeared in this team, who often had to attend long training associated with new or renewed components and controls. For scheduled shutdowns, for example in the summertime, when numerous customers also had their plant holidays, they have hired additional temporary personnel for maintenance, being not so familiar with the plant as their own employees.

In the last two years, numerous sensors have been mounted on the gear- and bearing housings and connected with one another. His rolling mill had entered a cooperation with the equipment supplier—a big plant manufacturer —aiming at making the maintenance more efficient to increase the operating time and to minimize operational disruptions—particularly the unplanned. Franks world of work was about to change as massively as ever before in his 30 years of service in the company. Previously he had been able to anticipate emerging bearing or transmission damage through observations of gears and bearings following unusual sounds with his experience—though it was difficult to hear these specific sounds in the operating noise. Thus he was able to prevent serious mechanical damage by timely shutdown. The result was often that the plant manufacturer had then suggested to reduce the service interval for all machine elements of this type. Then, when a rolling stand was exchanged and disassembled and the parts still looked quite "virgin", Frank always got annoyed that actually a lot of money was wasted.

The new sensors on the bearings and gears constantly measure mechanical vibrations on housings, shafts and foundations, as well as noise and temperatures. These were transmitted to the system supplier, who then tried—in a development project—to detect emerging disturbances due to changes in the patterns of vibrations and temperature gradients at different load situations and finally propose more specific and timely maintenance activities. On the other hand there were components which had indeed already exceeded the actual maintenance interval, which had not been necessary to be replaced. In the regular coordination meetings there were always surprises and insights that they would not have expected both for system manufacturers as well as for the rolling mill crew and for Frank as the responsible for the maintenance staff. Due to the emerging economic slowdown, the management now showed greater flexibility to allow disassembly of one or the other aggregate sometimes more often than absolutely necessary in order to bring together data of damage and data measured in the actually related production environment (dimensions, material etc.), and to interpret the sensor data correctly —in fact it was a huge learning process and effort. The mechanical work such as assembly and disassembly were paid by the equipment manufacturer, as they also expected to use these findings for other systems and plants. But the disassembling and assembling was much faster than before, since the technicians were equipped with data glasses. Thus the appropriate system components were mirrored three-dimensionally in their field of view, so they did not have constantly to look at large spread paper-plans or go for their computer to look at the CAD drawings. The majority were very young engineers, who had grown up with the computer. When Frank spoke to them, he realized, however, that they had little sense of which was the effect of a proposed inspection work on the production process and he had already rejected some suggested actions because a large part of the mill would have been affected. Here Frank's skill and experience came to effect, combining maintenance actions so skillfully, that the disruption of production was minimal. However, the production planning system of the entire stainless steel company was just in a phase of renewal and it should be possible in the next few years, to simulate production activities with unplanned and planned downtime from steel work to the rolling mill and on to heat treatment shops and thus significantly improve the production planning and controlling process. The young colleague always told him something about "Industry 4.0" and that the proposed maintenance measures would come from the computer system of the plant manufacturer in the future. In addition, certain parts would be produced in the future using 3D printers on site—something that Frank really could not imagine. The colleagues of the plant manufacturer had also told him that spare parts which could not be produced with this 3D printing technology, would be produced with lot size one at a manufacturer with a new "Manufacturing execution system" (MES), wherein the components had recorded their production plan and were able to "act autonomously"—choose the suitable and free machinery, automatically load the right tools and advance the manufactured component after the production automatically to the assembly line. That was "Industry 4.0", they told him. Next month he would visit this site with the plant manufacturer and thus have the opportunity to see all of that.

This scenario shows the following relevant characteristics of a subconscious mind of the organization steel mill in the narrower sense and the manufacturing industry as a sector or "organizational sphere":

- New technologies allow advanced organizational "perception" by sensors on the bearings and gears, etc. on the one hand and new ways to get necessary context sensitive information by augmented reality for maintenance technicians during disassembly and assembly on the other hand.
- Decision support based on predictive analysis from patterns recognized in the collection of millions of sensor data provide a better perception of what is going on and what is important.

- Risk of uncritical adoption of proposals from the decision support system and loss of "feeling" and intuition (*impact of individual measures on operating production activities is not considered by the computer-generated proposed maintenance activities*).
- The Information Infrastructure supports cross-company and multidisciplinary collaboration by removing technical hurdles and making collaboration and shared decision making easy (*coordination of maintenance and operations by simulation of specific events and optimization in order to provide decision proposals*).

4.2   The Central Component of the Organization: The Organizational Unit (OU)

The organizational unit (OU) is temporally unlimited—or at least limited until the next organizational change—consisting of several persons and/or OUs, which has a spokesman or leader who is responsible for the organization of this OU. System 1 and System 2 and the conscious as well as the subconscious mind are naturally to be seen more abstract in this structural element than the conscious and subconscious mind at the level of the basic element member, the individual person (Fig. 4.1).

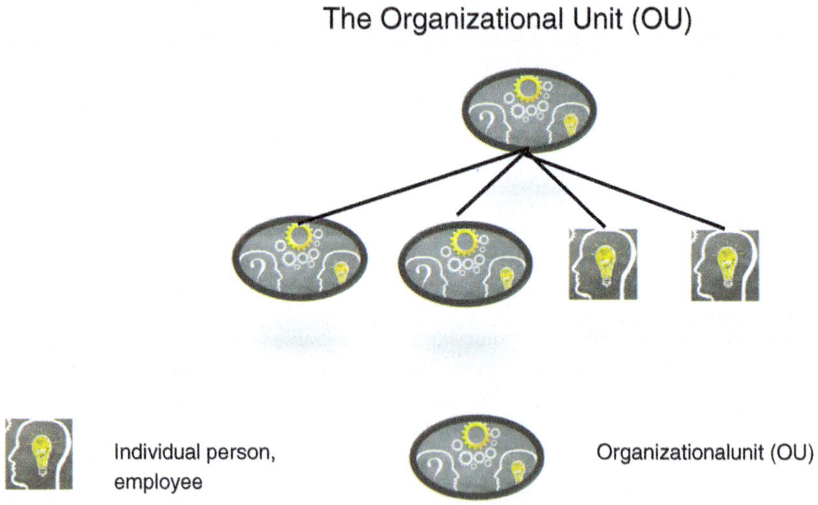

The Organizational Unit (OU)

Individual person, employee

Organizationalunit (OU)

Fig. 4.1 Structure of an Organizational Unit (OU), consisting of individuals and subordinated OUs

Decisions that are consciously taken at the level of an individual or a directly associated OU, may be perceived by the OU on the level above as consciously aware and consciously "incoming" (for example a note, a warning, etc.) and directly be incorporated into the consciously taken decision on that upper level. But these decisions may also feed directly into the system 1 (fast thinking) of this OU without taken notice of consciously, and thus affecting the upper level OU indirectly and unconsciously—being part of the subconscious mind of this upper level OU.

Example hospital group—*Leading a department, supporting collaboration, developing skills*
Sandra, the **chief physician** of the Department of Internal Medicine, has to cope with her process and quality responsibility, her ultimate medical responsibility for the patients as well as with the responsibility for the training of students, of physicians in training for General Practitioners (GP's—trainee doctors) and of physicians in training for becoming doctors of internal medicine. Thus she has to ensure that professional groups cooperate multidisciplinarily across medical disciplines in a way, that on the one hand the processes for patients are as patient centered and comfortable for the patient as possible and on the other hand organizational compliance (law on working hours, night shifts, emergency services, etc.) is ensured concerning her staff and the fullfillment of her department's task and mission. She leads on the one hand individual staff members, the medical assistance staff and care staff and on the other hand teams with their responsible leaders like the senior emergency physician with his/her team etc. The increasing specialization and resulting interdisciplinarity, like in oncology as well as other specialities, requires a high degree of co-operative and self-organizing organization—the often "military" command structures of the past seem obsolete. The organizational culture has changed. The chief doctor is on the one hand the highest medical instance and on the other hand teacher, assisting the learning of his team, and in particular enabler for good clinical practice and education. But Sandra as chief doctor is also responsible for the medical quality of the decision support systems. The department is highly dependent on the functioning of information technology infrastructure in the hospital and its manifold coupling to the information technology infrastructure of numerous partners. The tumor board with the structured and virtual case discussions—increasingly based on video conferencing—lead to good patient-related joint decision ("shared decision making"). For Sandra these tumor boards are a benchmark for future decision making in medically complex issues.

Previously, there was frequent tardiness and little discipline at the tumor board meetings, lack of mutual appreciation of specialists, lack of understanding for the previously often single-handedly taken treatment decisions in complex oncology cases etc. This has changed in recent years: the communication culture, organizational culture and consciousness of individuals have changed—for many staff members probably their subconscious has changed concerning their way of acting and interacting. The "subconscious mind" of their organizational unit has definitely changed so that there are far fewer conflicts than in the past. The "reflexes" of her department were conditioned more on cooperation than on demarcation now.

With concern, however, Sandra observes the increasingly uncritical acceptance of decisions proposed by the decision support system, which are derived from guidelines and adapted regulations provided by medical societies. This can be observed especially with the young colleagues still in training. Intuition and instinct of individuals are at risk of being and staying poor. Sandra is very concerned about the future quality of the department.

The new support for the discharge management in the form of statistical predictions of readmission Sandra considers as a valuable supplement because the statistical predictions can be interpreted as potential causalities. The staff is thus inspired to critical consideration of the entire patient-centered causality chain and that has a high learning effect. The instincts and the experience—and based upon these the intuition—of the staff members are sharpened. Although intuition can sometimes be a problematic counselor as Sandra is aware of. With extensive case studies as well as regular morbidity and mortality conferences Sandra believes to stay in control and minimize the risks of these new possibilities and to be able to develop the quality of medical care and education quality in the right direction. For Sandra, as the responsible manager of the OU, the easy to use access to relevant literature is especially valuable and they can incorporate it well in the training sessions as well as in the case studies etc.—a valuable contribution to the culture of evidence-based medicine. Particularly instructive are the decision support systems in medication where, especially with older, multi-morbid patients, valuable information on potential adverse effects and mutual influence of medication are provided. Prescribing so many drugs as never before is more and more in discussion and sometimes the consulting of pharmacists is required in very tricky cases.

The outcome quality parameters are now largely transparent between comparable organizational units of many hospitals. Considering the fact that, especially in Sandra's department, many old people are treated, these parameters are also "risk-adjusted" and provide a good comparison. This has greatly improved the quality awareness of all employees, so that all actions of the individual and especially the performance of the entire department in recent years has increased significantly. The consciously and often unconsciously-triggered actions have been greatly improved. Sandra and her staff are aware that they must be especially cautious with actions that are too subconsciously driven, that are no longer questioned and where neither the individual nor the organizational unit as a whole can take the time to sometimes stop, think and reflect. Therefore Sandra insists on the regular morbidity and mortality conferences and the commitment and participation of her medical staff.

The OU is with her information technology infrastructure interconnected in the sphere of internal information and knowledge sources that are actually part of the same infrastructure. But the OU is also connected to the sphere of external information and knowledge sources such as medical literature, drug interaction checks, relevant guidelines etc. The transparent representation of the quality parameters for example, represents the connection to the sphere of stakeholders, customers (patients) and competitors. The parameters and characteristics as well as the infrastructural integration of the different spheres of influence are described below.

This scenario shows the following relevant characteristics of the subconscious mind of the organization hospital in the narrow sense, as well as the health care industry as "organizational sphere":

- Facilitate and promote the culture of interdisciplinarity and cooperation with other departments. This will augment the skills of the team as well as of the individual staff members.
- Coupling to external knowledge such as literature databases, case descriptions etc. facilitates the use of this knowledge and motivates to take advantage of this. Evidence-based decision-making—and evidence-based medicine—are thus promoted.
- Use of the "data treasure" of the past (*predictions or Predictive Analysis in the area of discharge management*) as the organization's vast memory. These methods of predictive analysis are applicable in many areas of medical decision support,
- Ensure the working of the "System 2" through critical reflection of the decision proposals of application systems rather than uncritical adoption of the decision proposals by "System 1", so as to maintain the entire system in a permanent "learning mode" (*morbidity and mortality conferences*).
- Grant enough time to reflect on decision proposals. Do not use it predominantly as a productivity tool.

Example industrial company—*keeping production and maintenance in sync—with new decision making*
Ralph, the **managing director of the rolling mill** in his responsibility for the best possible outcome and efficient resource management of the mill feared nothing so much as unplanned disruptions and stoppages, since the complex and time-consuming re-scheduling of the production personnel was never really well mastered. On the other hand, the annual three-week stoppage in

the summer due to plant holidays was always a good opportunity to coordinate all planned maintenance, where—in case of doubt—components were often replaced, which would have held for several years more under certain circumstances. However, the statistics and the recommendations of the equipment suppliers recommended the "timely"—often too early—exchange. To reject such proposals could be interpreted as his wrong decision in the case of an unplanned disruption of production. Therefore, Ralph intuitively always took the cautious side and—only after lengthy budget discussions and risk considerations with management—he was willing to take a risk and then sometimes postponed a—according to the service interval recommendations —planned maintenance measure. In this case, a disruption due to this postponement would not be his fault but a joint management decision.

But now his man in charge of maintenance, **Frank**, was cautiously optimistic—with his innovative and inquisitive attitude despite his age—that the new maintenance strategies just being tested, which based on the continuous measuring by the various sensors, would be useful and the risk would be kept manageable. The new simulation and optimization algorithms and systems of production logistics from the steel mill to the rolling mill and on to the heat treatment shop and the final finishing of the products saved him the trouble of re-planning in case of an unplanned disruption. If all this would work as discussed, it could save a lot of trouble with his fellow managers of the heat treatment shop, the steel mill and the finishing line as well as the maintenance organization in the future. In particular, the maintenance organization could never provide enough staff when Ralph needed them. On the other hand he knew that experienced staff could perhaps be obsolete in the future with these systems in the background. Would the subconscious mind of his organization in the form of experience-guided decisions—though sometimes difficult to trace, mostly intuitive and often seemingly irrational—be replaced by an automatic decision system? Would that be a decision-making system, which would no longer be understandable? Would the supportive decision proposals soon be automatically adopted without humans having taken the time to reflect? That would mean the end of the learning organization, which they had so much emphasized ten years ago in order to initiate a continuous improvement process and to keep the organization alive.

This scenario shows the following relevant characteristics of a subconscious mind of the organization rolling mill in the narrow sense and the manufacturing industry as a sector or "organizational sphere":

- Sensors as expanded perception of the organization changed the approach of the management of the organizational unit and their interaction with the other organizational units thus leading to better mutual understanding.
- The simulation algorithms and decision proposals for rescheduling allow immediate rapid replanning (realtime) in maintenance-related shutdowns. Simulations are a way to communicate about complex issues and foster the learning of teams.
- The real-time simulation capabilities enable planners to enter in a "dialogue with the data" and to develop a new "sense" and "feeling" for the system rolling mill and its integration into the overall system.
- Staff only learns and develops as long as the system is used as a decision support tool and not as a decision automatism. In this case, the "Feeling" in the form of subconsciously available knowledge would be gradually get lost as a basis for quick intuitive decision proposals and ideas from humans. This lack of knowledge and sense would then be missing at specific decision-making situations and for the conscious reflection and validation of decision proposals.

The embedding of effective tools and systems of OUs into the fabric of the organization and the network of stakeholders will be described below. The tools and systems are essential elements of the perception and recognition as well as for memory, knowledge access and the decision-taking of an OU. Is it justified to speak about the conscious and subconscious mind? This still needs to be discussed finally later. But first, we will take a look at the working group as an essential element of an organization as well as an element of the model of the conscious and the subconscious mind of an organization.

4.3   Another Key Element of an Organization: The Working Group

Working groups can be seen as a special form of organizational entities, which are set up to deliver certain tasks by organizing the execution cross-organizational with the staff members assigned to this task but staying anchored in their respective organizational units.

Example hospital group—*Quality: confusion induced by indicators versus improved quality by "smart data"*
Margret works as **Quality Manager** of the hospital—being a trained physician, but no longer active in clinical operation. Her hospital is classified as a specialized hospital, which was in recent years constantly confronted with the performance indicators derived from the results of quality

measurements. Margret had to constantly "calm down" her chief physicians or delegated senior physicians—as Robert, who was representing the Department of Internal Medicine—in the quality assurance commission, because the already adjusted key quality indicators were not the best. The effectiveness and transparency of risk adjustment was often difficult to explain. So some departments have felt themselves discriminated and they already threatened that they would no longer have admitted seriously ill patients and patients with multiple diseases to their department for appendectomy because the higher risk for a possible death would result in quality indicators looking worse. This would be detrimental to the reputation of the Department. The "terror of numbers" induced by these quality indicators has been a danger to the working culture and induced fear of admitting critically ill hospital patients in need. The temptation to forward them to the central hospital, although the care and surgery—such as an appendectomy—could be provided within their hospital readily, was great. The clashes with the managers of the Central Hospital were accordingly unsettling. In most of them Margret was involved. Meanwhile in daily work "reflexes" and behavior patterns emerged that were not conducive to a good quality culture at all. "Has the subconscious mind of our organization already changed?" wondered Margret.

On the other hand there were also positive developments that were life-saving especially with elderly patients or preventing of severe disease. With reinforced and regular Point of Care Testing (POCT) medical parameters and vital signs taken by the nursing staff, the new monitoring system was able to achieve significant improvements by comparisons with historic similar cases and pattern changes of individual parameters and parameter combinations: Already hours before a doctor was able to detect the problematic development of an emerging sepsis, the system gave warnings so that timely action could be set. Thus, the "subconscious mind" of departments and hospitals could be improved with modern methods and automatically given early warnings.

However, it was expensive and without need to bring such a complex monitoring to all patients to use—young inexperienced doctors slightly inclined to including too many patients. Here the leadership of experienced senior consulting physicians and the physicians responsible for training was required.

The reports of Robert and the experiences of his department were impressive: The decision support functions for the attending physicians and nurses helped to optimize after-care management. Discharge management was improved with predictions on the expected readmission ("my patient") derived from cases of comparable patients from the "vast memory" of the hospital group ("like my patient"). These predictions were really helpful for affected older or severely ill patients. For Margret as quality manager these were certainly encouraging developments in her effort to contribute to constant improvement.

> The activities in the Quality Assurance Commission as permanent inter-departmental working group across their OU's had become much more powerful with the latest technological innovations and developments from Margret's point of view.

This scenario shows the following relevant characteristics of a subconscious mind of the organizations hospital and hospital group in the narrow sense as well as the health care industry as "organizational sphere" as a whole:

- Confusion induced by questionable indicators results in distorted perception: Indicators that are not comprehensible for the staff create uncertainty, mistrust and unrest and they disrupt the cooperation and working culture. This may result in individual decisions taken, which are potentially conflicting with the task of the organization (*avoiding the admission of elderly and multi-morbid patients*).
- The enhanced perception, based on patterns in the "Big Data" improve the effectiveness of the organization (*early warning of impending sepsis*).
- The subconscious mind of the organization—by the way of decision support and "dialogue with big data" has the potential to affect the quality of the organization significantly in a positive direction.
- Working groups, recruited from several organizational units that bring in different points of view reduce the risk of false, unconsciously triggered reflexes with counterproductive effects. Such teams and working groups are an essential element of the organizational learning process (*Quality assurance commission*).

Working groups usually have a limited lifespan like project teams until their mission is accomplished. In some cases they have an indefinite life span for example expert panels, where staff members from relevant OUs are delegated for the purpose of this working group (*such as the Quality Assurance Commission of the hospital*).

A classic example for a working group with limited lifespan is a project group, which brings together the delegates of interested OUs with their knowledge and expertise as well as the interests of the delegating OU for a temporary project.

> **Example industrial company—*project with partners—key question with outsourcing: which skills do I keep?***
> **Gabriel**, the **project manager** for the renovation and improvement of the maintenance of the rolling mill was commissioned by the management a few months ago to accompany the plant manufacturing company at the rolling mill, in order to monitor the improvements and to evaluate them regarding the use across the enterprise. He was expected to evaluate cost, benefit and implementation requirements and to develop an implementation plan with the accountable managers or their delegated engineers of the mills and workshops of the special steel manufacturer.

It was not easy to obtain—working with the system engineers from the plant manufacturing company—sufficient insight into their methods and algorithms. Their strategy obviously was to develop new maintenance services as part of their product portfolio in order to expand their business significantly at the customer site—in this case at the special steel manufacturer's rolling mill.

"What skills and influence would still remain with our company?" wondered Gabriel. Would there processes take hold that would automate the plant supplier's suggestions for maintenance jobs and thus enable him to get his—selected and profitable—maintenance orders directly from the system to himself? That would be not harmless for them as steel-producing organization: The actions taken would be significantly influenced by algorithms and automation in the background—and thus in the subconscious mind of their rolling mill—leading to decisions they would not be able to control nor maybe even consciously know about. And what about the many equipment suppliers in the heat treatment workshop and the large equipment supplier in the steel mill? Gabriel felt that he was chosen for a—even for his employer's future—very important task. The management also showed enormous interest in this project and again and again asked for information from the project.

This scenario shows the following relevant characteristics of the subconscious mind of the organization special steel producer in the narrow sense and the industry as a sector or "organizational sphere" as a whole:

- The work of a project group with its interdisciplinary expertise is well suited for conscious evaluation and design of what should remain in an organization's focus of awareness and what to settle in the subconscious mind of the organization. The events requiring the invocation of the "awareness" or the "vigilance of the organization" are defined here.
- The outsourcing of systemically important and significant planning and control processes to external partners (*for example outsourcing maintenance to the equipment suppliers and therefore giving away essential parts of the control of the maintenance*) affects the competence of your own organization essentially because the probability and risk is great that the affected outsourced area unwillingly leaves the sphere of the organization's own conscious decision making in its operating activities and—thus unintended and not controlled slips into the realm of the subconscious mind (*while it remains with the plant suppliers in their strategic, conscious focus in order to design the services offered and the future customer relationship as well as future revenue generation*).
- The subconscious mind has the potential to affect the quality of the organization significantly (*the efficiency of the maintenance of the rolling mill may outstrip if it is outsourced to the supplier because he can dispose of the latest technologies and sensor systems as well as pattern recognition algorithms*).

- Working groups staffed from several organizational units that interlink different points of view and reduce the risk of false, unknowingly triggered reflexes with counterproductive effects are an essential element of organizational learning process.

This organizational form of temporary or permanent working groups can be used as a deliberate measure against intuitive reaction patterns of inertia—controlled or triggered by the system 1—which follow the interests of single OU's only. The working groups lead to structured conscious solution finding for the organization as a whole. Working groups and mixed expert panels also promote the conscious dealing with overarching themes—of well established reaction patterns and behavioral patterns. Commissions, like data privacy commissions, platforms for enhancement and improvement, quality circles etc. can—if well-moderated and provided with clear objectives—break and improve deeply rooted longtime unquestioned procedures. But beware of pure "meetings for talking".

4.4   The Organization as a Whole

The organization as a whole is defined by its purpose—be it in the form of articles of association, bylaws, contractual agreements with real or virtual organizations or even merely informal mutual understanding based on shared visions and values.

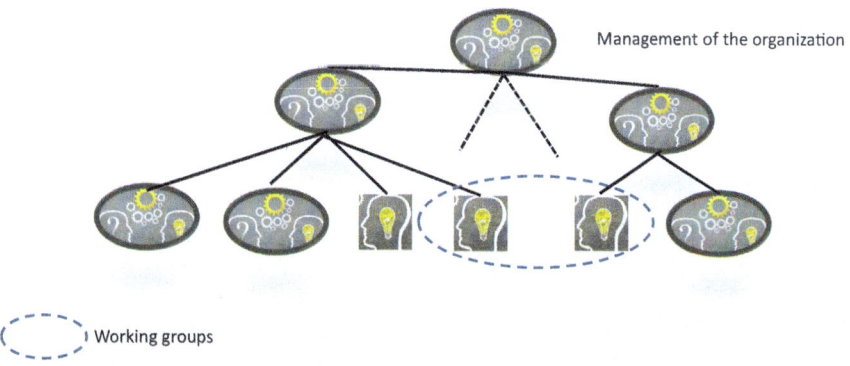

Fig. 4.2 The organization as a whole

The organization is responsible for the fulfillment of that purpose, representing itself in its external dimension and being part of a social construct itself, be it on local, on national or on international level.

In Fig. 4.2 an organization is shown schematically—its organizational units (OU's) and its employees, with cross-OU working groups and a management or a governing body on top. The organization is "woven into a fabric" with its stakeholders and anchored therein, trying to accomplish its mission and follow its strategies - with one rising challenge being adaptiveness for actual and future questions often posed by dynamic and exponential developments in the technology field.

Example hospital group—*increased transparency of patient processes pushes the learning organization*
Paul is the **CEO** of the hospital group, which also includes the hospital with the department of Sandra. A big problem is delivering patient-centered care with the best possible coordination between primary health care centers and standard hospitals, the main hospitals and the central university hospital. Some are hospitals of his hospital group and some are hospitals belonging to other organizations. Certain services are provided in several hospitals, thus also being in internal competition. Certain medical services—especially specialty services such as cardiac surgery or neurosurgery—are provided only in the central hospital, the university hospital.

Addressing the trade-offs in terms of performance with good profit margins, good utilization, medically interesting cases at the departmental and hospital level on the one hand and patient-related quality of care on the other hand, Paul's experience definitely is that things do not always happen for the benefit of the patient only. Employees follow—within their margin of discretion—often quantifiable targets that are often associated with financial incentives—especially with executives. "Patient-centered care" as postulated in the corporate vision sometimes comes a little short with the numerous decisions taken for patients regarding admission, transfer, discharge, reappointment etc. The patterns of motivation of individuals, their points of interest, their personal networks, especially within the medical profession, as well as personal preferences, social awareness, specific ethical attitude etc. form a hard recognizable subconscious network—being a part of the "subconscious mind" of his organization. In recent years it has become increasingly easier—facing the nationwide outcome quality indicators—to avoid, that cases were treated at local standard hospitals, which actually belonged to a central hospital. Earlier the individual physician or chief physician with his professional record from the University Hospital had wanted to treat such cases and often in fact treated them. The new transparency led to more discipline in complying with the—to each hospital—allocated spectrum of medical services. On the other hand, telemedical consulting with the central specialists also allowed to handle more sophisticated diagnoses and therapy locally with good quality and patient safety. But these telemedical services were poorly paid and therefore not popular, although thus high quality care close to home with the telemedical involvement of specialists was quite feasible.

The new methods with Big Data now allowed "interactive dialogues with the data" supported by graphic visualizations of patient-related events and intuitively easy to use applications. Patient histories of similar cases now could be statistically compared with the new algorithms and technologies and the now affordable powerful computer resources. Thus also statistically-based predictions could be provided by decision support systems. Massive deviations from good clinical practice could be detected by visualizing different treatment patterns and clinical practice as well as "patient processes" through the system. They could be analyzed related to predicted and average outcome and thus—in case of positive deviations—the standard and established practice could be improved or—in case of negative deviations causing less quality—they could be avoided in the future. These were significant contributions to knowledge management and for Paul it seemed to be a milestone on the way to his hospital group becoming a truly learning and—finally adaptive—organization. In discharge management in the department of Internal Medicine managed by Sandra as chief physician that already worked.

Paul was confident that false incentives and motivations as well as attitudes conflicting to the goal of patient-related quality would become transparent in their outcome and thus be objectively discussed. So subconsciously guided behaviors became conscious and the subconscious mind of his organization was changing in the right direction.

This scenario shows the following relevant characteristics of a subconscious mind of the organization hospital and hospital group in the narrow sense as well as the health care industry or "organizational sphere" as a whole:

- Incentives and attitudes—often influenced by corporate culture as well as the respective acting personalities—mostly unconsciously determine the behavior and the decisions and respective actions taken. So the setting of decision-making is partly determined by the subconscious mind of the organization as a whole.
- The proper design of the corporate culture, the strategic direction and the selection of key personnel as well as the appropriate decision support tools (for example, discharge management) can lead to well-functioning organizations with functioning interaction between the subconscious mind and the conscious mind of the organization and thus lead to efficient business processes.
- The management and control effort for permanent control of these largely automated ongoing decision-making processes can be kept low and the span of control can be kept high because the management load for top management is relatively low when the subconscious mind of the organization is properly designed.

- Otherwise, actions of intervention need to be continually set, which interfere with the efficiency of the processes and bring "stress" to the organization.
- The complexity of management and control can be kept manageable with methods of Visual Analytics and Big Data (*detection of deviations in the dialogue with the data*).

Some aspects of the operation of the subconscious mind and its ability to influence become evident in this scenario and its analysis. Management tools and infrastructure to manage this complex field of conscious and subconscious aspects, are urgently needed. It is a challenge for the managers when wanted and unwanted effects of these conscious and subconscious aspects affect each other and when these effects are not evident at first glance.

Example industrial company—*new business model—what is our core competence—what skills are needed?*
Martha is one of the few female **top managers** in the steel industry and manages the special steel company, in which we have already met the operations manager of the rolling mill, **Ralph**, the maintenance engineer **Frank** and the project manager **Gabriel**. The operations managers of the steel mill, the heat treatment workshops and the finishing operations as well as the manager of the maintenance organization are men in the 50 s, who had already experienced heavy corporate crises. Martha's job as managing director with experience in marketing and management of distribution channels was to bring the distribution channels and the needs of the market in resonance with the possibilities of efficient production processes. The patterns in the production in this industry were by tradition technologically aligned to the throughput of high tonnage on large aggregates. But the market for special steel products increasingly demanded small lot sizes of specific tool steel and other special steels—a very flexible "apothecary shop", some remarked. This resulted in high inventory and capital costs as well as frequent devaluations of inventory that heavily weighed on the bottom line earnings. Over the past decade numerous investments were made in order to become more flexible for example the feature that parts of the rolling mill were able to produce, when at the same time a line of the mill was converted for other dimensions of rolled products or was just being repaired. In the steelworks the steel-making furnace and the casting system was equipped for producing smaller batches and smaller block sizes so that semi-finished products could also be produced efficiently. This had required a partly integration with the heavy section mill in order to reduce energy demand of reheating furnaces.

The planning and control systems were just in renovation. Simulation, optimization, decentralized autonomous control, Industry 4.0 etc. were the key words that Martha heard again and again. From engineering and automotive industry Martha already got to know some interesting approaches and first results. The steel industry was something else—no assembly and no merging of parts, but an increasing demand for small delivery volumes in the respective quality that had to be melted in large quantities in the steelworks, then were formed in the rolling mill in different dimensions and then were finalized to customs order in the heat treatment workshops, then in the finishing lines and finally prepared for shipping. On the one hand enormous flexibility was required—on the other hand there was enormous cost pressure in operations.

In maintenance she had launched and patronized a project with the young ambitious operation assistant Gabriel as project manager. They just tried something new: Massive use of sensors as a sort of "Enhanced Perception", but which was only—up to now accessible and usable by the plant suppliers. Martha thought about whether it would be wise to outsource the maintenance of the rolling mill to the plant manufacturing company and pay them for production availability and uptime. Unplanned shutdowns would then be penalized. The plant manufacturing company would then buy back the rolling mill at fair value, Martha would quite freely get to capital in order to thereby open up new markets.

She already figured out the discussions with their managers about this entirely new business model. For most of them such ideas were inconceivable: To produce on plants that did not belong to the company and they even would not maintain these machinery themselves. The maintenance staff on site would then be working mostly for the plant manufacturer. Would it really be wise to do something like this? Would the company be losing core competencies substantially? Or would her company and her staff be able to build up knowledge for the sensor systems, and especially knowledge for the early identification of necessary maintenance measures based on the vibration patterns, temperature profiles, etc. of gears and bearings etc. in the aggregates at different load situations—thus avoiding unnecessary preventive maintenance as suggested by purely interval based maintenance strategies? But the plant manufacturer would have new, different and more enhanced possibilities for comparison. To build a rolling mill finally was his core competency and he did and does it at many sites and at many companies in the world, also in many companies and sites producing special steel products—the competitors to the company managed by Martha. In fact by this way the plant manufacturer would have access to their production processes in the rolling mill in detail—one of their core competency in their markets. In any case, thought patterns of her employees and the conscious and unconscious behavior of her organization would significantly change—it would have to change—with these new technological possibilities.

This scenario shows the following relevant characteristics of the subconscious mind of the organization special steel production company in the narrow sense and the industry as a sector or "organizational sphere" as a whole:

- The flexibility of the new possibilities of simulations, of the Internet of Things (IoT), semi-autonomous systems and—in a comprehensive view—of Industry 4.0 changes the possibilities of the company. The management challenge imposed by the new flexibilities is to handle increased complexity. This leads to an increase in the number of individual decisions to be taken that can only be handled when this decision making was implemented to be run quickly and automatically in the "subconscious mind of the organization".
- This will require efficient tools for control and recognition of decision situations, which require the active involvement of management. It is important to establish "if and when" to switch from unconscious control to conscious control ("System 1 to System 2" following Kahneman) and it is important to recognize early on, when the system is running out of control and problematic trends evolve.
- Which skills do I need in future? What are my core competencies which I will need in future. Will I have to change my business model significantly?

These considerations make evident how necessary it can be to change thought patterns and attitudes of management as well of the staff—and finally of organizational units and the organization as a whole—in the light of new technological possibilities in the fields of "extended perception" and "new decision making" by the organization. Which paradigms do I have to adjust? These changed or new paradigms would become part of a modified "subconscious mind" of the organization—favorably to be designed and shaped deliberately and not "happening" without control in case of a possible crisis of my organization. The challenge is to be adaptive in a controlled way to changing markets and constraints.

4.5   Cascading of the Conscious Mind and the Subconscious Mind in an Organization

According to the organizational levels of a specific organization the respective conscious minds and subconscious minds are cascaded in self-similarity, each with its specific characteristics:

- **From staff members**
  with their intuitively taken decisions (System 1 by Kahneman) and their conscious decisions influenced by their subconscious mind, which in turn is influenced by the tools of "extended perception", "extended recognition" and "decision support" provided by the organization: *Senior physician Robert would probably not have taken into account the osteoporosis risk in the standard discharge process. Only by the extended perception in the form of the easily accessible health record and the "extended recognition" in the form of the prediction of readmission probability he did recognize the significant risk of osteoporosis.*

- **on to the organizational units**

  in their respective level within the organization with their formation of a "collective subconscious mind" formed by its patterns of behavior and social as well as professional interaction for example of young employees of an OU, who have been working with the decision-support systems from the beginning of their careers. So these tools with their algorithms behind the given decision proposals have already emerged to be a part of the "subconscious mind" of their OU: *Chief physician Sandra wants to off set the danger of uncritical adoption of decision support proposals from the system by her young staff members by regular case conferences and conferences on mortality and morbidity in order to actively form the subconscious mind of her OU.* The decisions taken by staff members or OU's (the respective OU is responsible for) are for the management of the OU already part of the subconscious mind of its OU. The formal and informal rules what has to be decided by staff members and what has to be decided in the group or OU shape the subconscious mind of the respective OU.

- **to the organization as a whole**

  and its subconscious patterns of behavior and thinking—resulting from all staff members and OUs, from available and provided tools as well as from organizational culture and organizational design: *Martha as general manager is faced with the almost unshakable opinion of her managers that the facilities should be strictly owned by the company and maintenance expertise should be mainly with the company's employees—an organizational "mindset" that would be hard to change or even adapt.* For the organization as a whole and its top management the decisions taken in the OUs, working groups and in business processes are part of the top management's subconscious mind of the organization. Only decisions being an inherent task of the top management level or decisions delegated back from lower levels become aware and become part of the top management's conscious mind (Fig. 4.3).

Cascading Subconscious Minds
form the Subconscious Mind of your Organization

Fig. 4.3 Cascading Subconscious Minds form the Subconscious Mind of your Organization

We can clearly recognize a cascading of self-similar behaviour and thought patterns, which is not surprising, since an organization is composed of people who in their assessment patterns and decision making patterns are—following Kahneman —subject to thinking fast (the subconsciously triggered System 1) and thinking slow (the conscious deliberate thinking system 2 also influenced by the subconscious).

The willingness to turn on the laborious and exhausting system 2, and thus to enable critical reflection challenges the structure and the management of an organization or an organizational unit enormously. Of course the system 1—the thinking fast—is in humans a faster process than in the "system 1" of an organizational unit or even an entire organization. However, the characteristic to follow a deeply rooted pattern of behavior with little effort is also valid for an OU and organizations and can certainly be regarded as an expression of the "thinking fast" of an OU or organization as a whole.

Fundamental new approaches (such as the envisaged outsourcing of maintenance) require to change the subconscious mind of the organizational units, otherwise the OU's may remain in the mode of the system 1 for a longer time and the mode "System 2" with its active reflection and willingness to change may be avoided as long as possible (possibly resistance). Often "system 2" can be mobilized only by the organization as a whole—by the top management.

Particularly large complex organizations with high claim of central control often do not have the strength to get out of their "system-1" mode and switch to the "system-2" mode in order to change the behavioral patterns practiced. Often a veritable crisis must occur to switch "5 min to twelve" into an innovative and challenging system-2 mode and mobilize the resources accordingly.

The higher we come considering the organizational cascade from the employees (the individual acts—thinking fast (System 1)—in fractions of a second) to the organization as a whole, the longer it takes until the "organizational intuitive" mode —controlled by the organizationally subconscious System-1—is left and the more strenuous and exhausting the change process takes place until the organizationally conscious system 2 mode finds its way and reflexive resistance disappears. It is a veritable challenge for good leaders to accelerate this change process and to get a modified subconscious mind of their organization working.

The self-similarity—albeit at various time scales—between the respective stages of organizational cascade seems evident.

On the other hand a well adjusted "subconscious mind" has the potential to allow a relatively effortless and efficient system 1 mode and a good and efficient organization through timely and accurate decision taking. Permanent action in the system-2 mode would be too uncomfortable and inefficient. A driver of a car could not drive a car if he would constantly act in the resource-consuming system 2 mode. Living our lives as humans requires this dual-mode system. Evolution has decided this way. The respective metaphoric view on organizations thus clearly seems to make sense.

Working groups, project groups, process management teams and the managers have the task to constantly pay attention to the good "calibration of their subconscious mind" in the respective level in the cascade by temporarily consciously putting themselves in the System-2 mode and to reflect their actions and the organizational behaviour of their organizational unit or their organization as a whole and thus seek a continuous improvement process. The possibilities for the design and the shaping of the subconscious mind are discussed in Chap. 5 of this book.

Organizations always have a purpose when they interact with their environment. It is in this interaction in their costumer relationship, in supply chains etc. that they become effective and generate impact. The model of the subconscious mind of an organization must therefore pay particular attention to this connectedness to the environment of an organization. That is the focus of the following section regarding the modelling of the subconscious mind of organizations in a metaphoric view derived from how humans perceive, recognize, conclude, decide and act.

4.6 The Environment of the Organization and How the Organization, the Organizational Units and Their Staff Are Embedded in It

The environment of an organization can be structured in 3 spheres (see Fig. 4.4):

- The sphere of stakeholders and customers,
- The sphere of competitors and the market and
- The sphere of external sources of information and knowledge that are partially inspired from the other two spheres.

The Environment of Organisations

Fig. 4.4 The environment of an organization can be structured in 3 spheres

The **sphere of stakeholders and customers**—as part of the market—includes the customers of the organization and the stakeholders, such as owners and the public, councils, boards, general assemblies, platforms and forums, which significantly affect the organization and its control, and where the head of the organization is responsible to and accountable. The expectations of this sphere have considerable influence on the strategy and the "internal constitution" of the organization as well as on the attitude of its staff. This internal constitution of the organization and the attitude of its staff affect the intuitive action—the working of the "subconscious mind"—a great deal. It controls the routine activity of the system 1 in order to meet the goals of the organization and it controls when the system 2 comes on the scene and consciously and reflective acts, giving constant feedback on the system 1 in order to contribute to the learning organization. Bylaws, partnership agreements, internal control systems, compliance rules and numerous other regulations formalize this process.

Slightly less structured is the link to the interested public in the way of corporate communications and public relations. Crises, unwelcome publicity events, but also the communication of successes feed back to the members of the organization in the way of media such as newspapers, journals, radio and television, thus influencing attitudes and the subconscious minds of employees and organizational units. These act on the organization similar to how humans are affected by praise and blame, reward and punishment, attention and non-attention thus conditioning thinking fast (System 1) and influencing behaviour of individuals and the organization.

Costumers are on the one hand in general part of this interested public, on the other hand they require specific focus of the organization and its costumer relationship management. At least they mostly pay for services and products. That is what the organization lives from. But that is not so clear in all settings. There are also business models like in healthcare when the patient as costumer only pays indirectly for the service—via his health insurance.

Suppliers as partners in the value chain also can be viewed also part of this stakeholder sphere.

The **sphere of competitors and the market** covers the opportunities and threats arising from the representation and action of market participants, from economic and environmental analytics and forecasts as well as from technological developments and trends. These expectations, trends and forecasts are perceived in very deliberate and analytical processes of the organization. Opportunities and in particular the dangers of substitution, displacement, fading of demand etc. can—if constantly and badly communicated over longer periods—impact the attitude of employees, organizational units and the organization as a whole in the way of fear from loss and change. These fears and attitudes find their way into the subconscious mind of the organization and accumulate there. But also opportunities, a climate of innovation and entrepreneurship in the respective business area have the potential to influence the subconscious mind of the respective organization significantly. This leads to the third sphere of influence:

The **sphere of external sources of information and knowledge** is often given only little attention especially at very self-centered organizations. Sometimes the resulting opportunities and effects are not recognized as such. The way to make use of these resources and their connection to the organization, the OUs and the staff as well as the role model lived by the managers in dealing with these resources are part of the corporate culture and penetrate into the subconscious mind of the organization. Curiosity and the search for and exploitation of opportunities as a fundamental value cannot be decreed, but only be stimulated, be exemplified and developed. This sphere is dominated like no other through innovation and change. This is the case with Web 1.0 with its search engine as new source of information—as before journals, market studies etc. were sources of external knowledge. This is also the case with Web 2.0 and its social media like Facebook, Twitter etc. as models for tools of participation for employees, organizations, customers and other stakeholders. Open Data—the provision of public data—offers new opportunities in numerous entrepreneurial and organizational areas. Even literature databases are worth mentioning here. The even closer integration with the environment and extension of perception by the Web 3.0 with its tools of Augmented Reality and the Internet of Things open up new, still largely unexplored possibilities: Opportunities for Innovation for agile organizations—a threat to the non-movers and late adaptors.

The connection of the organizational units and their staff directly goes to these "external" spheres. In any case, they are also connected to the sphere of internal sources of information and knowledge, that are in turn part of the technological infrastructure of the organization (see Fig. 4.5).

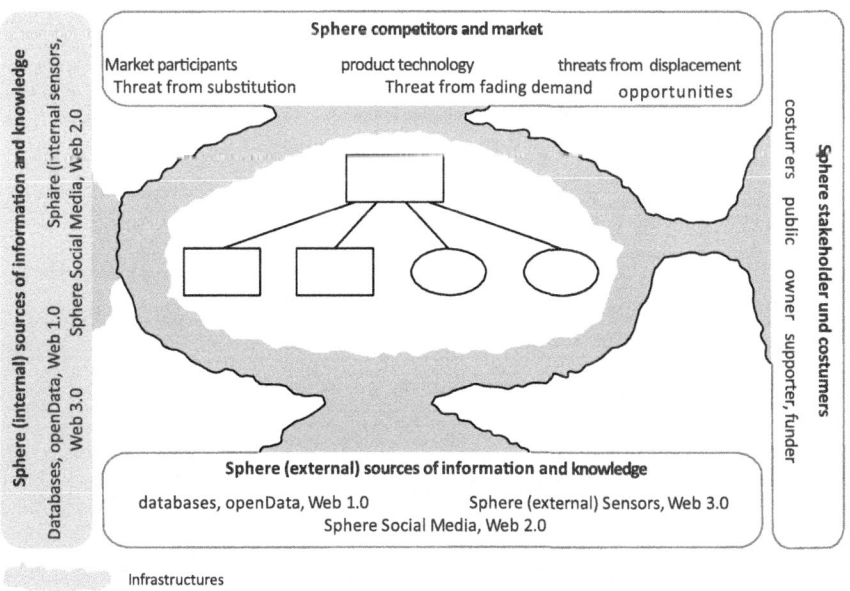

Fig. 4.5 Embedding an organizational unit (OU) in the external and internal spheres of influence

These internal sources of information and knowledge include databases and applications of all types, the Content Management System, Intranet, Process Management systems or special knowledge management applications—as well as e-mail accounts. The performance of the search engines and their usability and the accuracy of their results which should ideally be constantly improved for the respective target group based on learning algorithms, is of particular importance, since data and information are increasingly unstructured—but when stored they are already often "tagged". This world is more easily manageable for "digital natives" than for "digital immigrants" who "professionally matured" with paper.

The embedding of the organization as a whole into the spheres of influence of its environment is outlined in Fig. 4.6.

When you rethink your organization try to shape the subconscious and conscious mind of your organization in a way that all levels of your organization are linked and embedded in the outside world, in the environment of your organization. Open up and encourage self motivation at all levels. Do not be afraid too much of lack of control, lack of focus and lack of efficiency. Empower your organization to embed itself in its surroundings and its business area.

The organization as a whole is embedded in the information technology infrastructure as part of the subconscious mind of the organization. On the other hand, the information technology infrastructure and in particular the organizational ("soft") infrastructure—such as integrated management systems—form the interface between subconscious mind and conscious mind of an organization. The

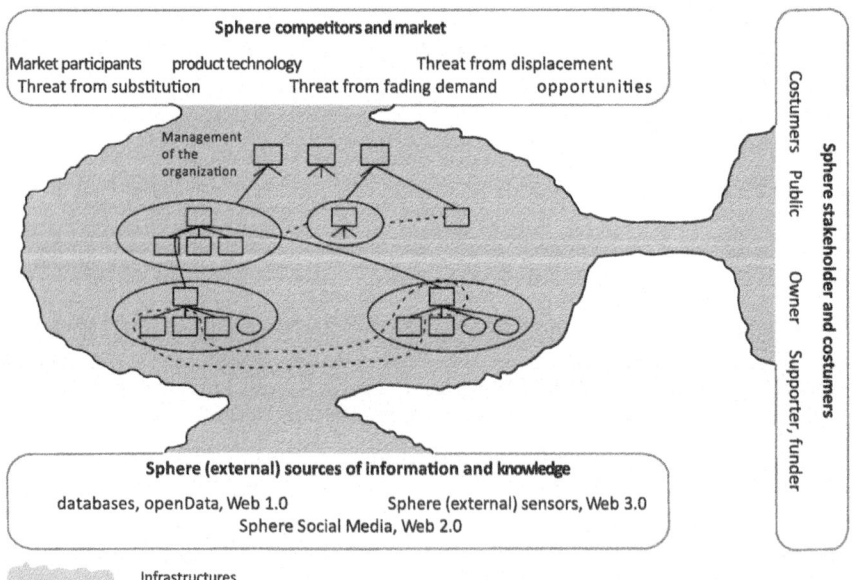

Fig. 4.6 Embedding of the organization as a whole in the external and internal spheres of influence

instruments and the elements of an integrated management system like ICS (internal control system), risk management system, controlling, process management system etc. are the triggers that detect deviations, detect change and improvement, take deliberate steps toward reform, recognize need for involvement of the senior management etc.

4.7   The Infrastructure of an Organization

Probably the most essential parts of the infrastructure of an organization are the information and communication systems on the one hand—including internal information and knowledge resources together with the connection of the organization to external resource of information and knowledge—and the management system itself on the other hand. In Sect. 3.1 infrastructures were treated in a broad meaning especially regarding "soft" infrastructures (see Sect. 3.1.1) and ICT infrastructure (see Sect. 3.1.3).

Office and production infrastructure—dealt with in Sect. 3.1.2 as a "technical infrastructure"—are also important elements of an organization's infrastructure as far as the working conditions and their effect on organizational behaviour of the employees and workers are concerned. To be mentioned in this context are the office space and the question of how they are suitable for the style of work and the labor needs of the organization and how they reflect the organization's habits, culture and style: lot or a little traveling, telecommuting options and home office, the significance of personal contacts versus video contacts, cubicle or single office, availability of extensive additional meeting options etc. These affect the informal communication in the organization significantly. Not least, it is thereby made easier or more difficult to make the "tacit knowledge" of the organization available thus bringing it closer to the decision-making process and thus enriching the subconscious mind of the organization.

> **Example Hospital Group**—*patient-centered healthcare in a network of providers—EHR and Quickview*
> **Paul as CEO** of the Hospital Group is aware that his hospital's use of the information technology infrastructure is about to extend their perception. The findings and medical reports of the other hospitals of the hospital group are integrated into the decision-making process with regard to diagnosis and therapy of the respective individual patient.
>
> The electronic health record (EHR)—a system outside his hospitals, run by public organizations—provides the preliminary findings for each patient from all relevant healthcare providers—unless the patient has opted out. For **Sandra, the chief physician** of the Department of Internal Medicine, this was indeed progress, but the need to take a look at the EHR of the patient was added on top of an already heavy workload for her physicians. With

chronically ill patients they had to scan long lists of medical documents looking for relevant prior information, but it was hard to get an overview of the relevant information. For one year, they now had experience with a helpful context-sensitive function as feature of their hospital information system that displayed to the doctor keywords—relevant for their respective working situation—as well as relevant documents and information in a form, that allowed experienced physicians to intuitively decide, whether they should go deeper into the medical history or not. They called this function "Quickview". The patient interview thus could be continued quickly and efficiently. Inexperienced doctors tended to only look at the results from actual laboratory tests etc. and hesitated to dive into the—now available entire medical history—although it would have been necessary sometimes. They were often reluctant to ask and talk to the patient in order to get the relevance, the plausibility and the completeness checked. Some patients also showed their astonishment not having even been asked about their "medical history". The new tool "Quickview" was about to improve the quality of work in Sandra's department.

Sandra had learned that the tools generated different patterns of how users were dealing with these tools. That was why regular reflection of the way to work with these tools in feedback and discussion meetings were important to Sandra, because that formed the quality of subconscious action and fast conclusions significantly over time—be it with individual or with group decisions. The high quality of the fast and intuitive—almost automated—decisions and actions derived from the organizational subconscious mind of her department and the subconscious of her doctors had become an important element of her department's overall quality. The quality of their interconnectedness in the healthcare system as a whole had improved significantly—to the benefit of their patients making new ways of patient centered care feasible.

This scenario shows how the technological infrastructure and its tools become—accompanied by extensive communication and learning opportunity in meetings and conversations—part of the subconscious mind of the organization and a defining quality factor.

Example hospital group—*good corporate governance* via *an integrated management system* **Paul, the CEO** of the hospital group, was conscious about the difficulty of his task when he was appointed to this position. From his previous position as senior consultant in an international consulting firm he knew many health organizations. Before that he had already met some health organizations as Executive Assistant of a large university hospital complex. The chief physicians and university professors were responsible in their respective department and were also very influential as opinion leaders, with their numerous cross-connections to decision makers in politics and to other stakeholders. On the other hand Paul had—with his management team —the final responsibility also in the sense of legal liability. The budget constraints—given the scarce financial resources of the public sector—were felt as a threat by many fellow managers and staff members.

Paul had succeeded in recent years, to formulate the policy and the strategic orientation of the Group in dialogue with the political leaders and other stakeholders and to anchor them with the correspondingly agreed management objectives from year to year. Now he was going to merge— together with his employees in the corporate headquarters—the various elements and management tools such as risk management system, information management, quality management, internal control system, etc. and thus to support the strategic and operational focus, on the one hand preventing that the thus formed integrated management system not becoming a bureaucratic monster and on the other hand securing a good governance. It was his goal as CEO to set out important aspects, such as the mission (above all the patient orientation), the strategic directions and objectives, the risk awareness of the most relevant topics, the compliance that they had already formulated in a Code of Conduct, the Awareness in the field of data protection and data privacy etc.—He wanted to anchor this in the "subconscious mind" of his organization, so that his managers and employees would automatically incorporate these aspects in their daily operational decisions. His aim was to anchor this in the form of a comprehensive tool for his hospital group and its business—an integrated management system as guiding and as compact as possible.

Management tools and methods for the internal processes, but also for the connection to the external spheres are part of this infrastructure as well as an essential part of the subconscious mind of an organization ("soft" infrastructure—see Sect. 3.1.1).

Example industrial company—*Enabling better inhouse maintenance versus outsourcing—a strategic decision* **Martha**, the **managing director** of the special steel company, noted a steady increase in the external labor costs in maintenance, without the personnel costs reduced in their own maintenance staff. The maintenance manager and the operational managers of the different workshops argued that the increasing costs for remote maintenance was due to the—now also in the steel industry—ever higher degree of automation with a lot more automation components and measurement points and sensors in the plant. In case of problems or faults—warning signs blinking frequently without the operation being in real danger of a massive standstill—the supplier obviously had to be increasingly involved.

Unfortunately, there was still too little standardization in the components of the systems. If there were standardized control and visualization platforms they would not need as many employees holding the specific knowledge for those many different platforms. Martha granted her operations managers half a year to prepare a fundamental decision regarding the future maintenance strategy.

The alternatives were: mainly inhouse maintenance or outsourced maintenance by specialist companies on their respective platforms. But this outsourcing path would require consequent reduction of plant personnel. In case of keeping maintenance inhouse a clear reduction of costs for outsourced maintenance was required by standardizing control and system visualization tools and platforms, and thus enabling inhouse personnel to control the plant maintenance better. For this alternative—according to the head of the IT department—there were already some encouraging signs visible.

The structure, functionality and standardization of the connection of an organization and its assets to external partners and suppliers are of strategic importance—not only for the cost structures but also for the skills of the organization. This decisively influences how quick and "automated" the organization can react to problems and fix them or how slow and complex that goes. Does a problem of a certain severity need the acknowledgment or even an intervention from the upper management level? Does it advance or have to advance into the awareness of corporate management or should it stay in a well conditioned subconscious mind of the organizational unit or workgroup in charge—from the perspective of corporate management? Structure, functionality and standardization of the connection of an organization to its environment are therefore strategic decisions which pose the question of the core capabilities an organization needs—and thus to the subconscious mind of the organization. This structure, the functionality and standardization of the connection of the company to its environment are therefore also to be seen as infrastructural aspects of an organization and to be taken into consideration.

Example Retail Company—*costumer loyalty in local retail business in competition with Amazon* **etc.** For **Mark** as a **marketing manager** in a retail company of consumer electronics and household appliances industry, the last professional years were really stormy. After their sales outlets—with the consulting service for the costumers offered there—competed as a sales channel increasingly with pure online sales channels like Amazon, etc., he was called upon to make better use of the relationship of his company to the customers and to create something like greater customer loyalty. However, prices had to compete with the online sellers who increasingly succeeded to deliver any product in two or three days with their sophisticated logistics chain. For Mark it was evident that advice and consulting for costumers by experts in business and the use of storage with the "takeaway option" for the costumer—compared to online distribution channels of Amazon etc.—could be a way to stay in business with locations near the customer. However, he actually hardly knew his customers—compared to Amazon, which, as he knew him from his own experience, made suggestions for other books to buy after he ordered a book, because Amazon etc. obviously already knew much about him. He already knew well from the last marketing congress that he attended, certain keywords and concepts, such as "Smart Commerce" and "user experience". He did not have much time to submit proposals to the management how to improve and open up opportunities for future sustainable development of their retail company.

Using the example of the consumer goods industry and consumer electronics and household appliances industry with their market access, the importance of infrastructure and systems for access to the market and to the end-customer can be discussed best. Probably each of us is confronted with it and had their own user experience with

- incompetent sellers, where you searched for advice in vain,
- with hotlines, where one was further connected after a long wait and
- annoying calls from sales people, who wanted to sell something.

Discussions with friends and acquaintances show then, that the experiences are different and specific contact forms and procedures of the providers are perceived differently: The one expects complete anonymity: no name, no email address, no phone number. The other wants the providers to give advice and then goes on bargain hunting and price optimization into the Internet.

Networked customers using mobile devices and social media technologies in a variety of channels, automatically generate vast amounts of real-time data. It is necessary to convert this enormous amount of data into meaningful information, to enhance customer experience and to create increased customer loyalty and so on.

Customers are well informed, they are often linked continuously and exchange with each other through social media. They shape opinions in a manner which is beyond control for the organization. The subconscious mind of the organization including its links to the costumers thus is largely defined from the market and by technologies. What can be influenced by the respective organization by shaping its subconscious mind?

The customer expects a brand-specific costumer experience that is as far as possible tailored to him and consistent in itself: on the website, in e-mail marketing, in advertising, in the search engines, in the call center, in the social media fora and in the aftersales services and assistance. According to Forrester Research Inc., the online sales in the US 2012–2017 are expected to increase 60% to $370 billion (see

Forester Research 2013). The increase in web-enabled "smart appliances"—from cars to television—and in point-of-sale transactions on mobile devices will offer new opportunities for the development of closer relations with the customers.

The aim of such infrastructures and applications for customer loyalty is to help customers with the right processes and technology to find the right product/service at the desired price and to buy it as well as to obtain necessary information about it. These costumer loyalty systems want to advice for the desired product or service and to motivate the satisfied costumer to recommend the brand to others—regardless of location and the used distribution channel.

For this purpose it is necessary to align the organization/company/employees to view each customer as an individual personality. This requires

- creating a culture of absolute customer focus and
- the proper instrumentation at the essential points of contact (for example, call center, CRM—Costumer Relationship Management systems),
- the creation of an illustrative "image of each customer" (including the link with their social media data),
- the generation of meaningful information directly at the customer contact: predictive information ("What else could I offer them?"), as well as historical costumer and costumer transactions data.

Altogether that allows—in combination with other marketing tools—the creation of an authentic and consistent brand culture.

This leads to a completely different and new connectedness to the sphere of competition and market—compared with today's connections and access of an average organization, be it a profit or a non-profit organization.

The "conscious mind" of the organization with its truly conscious and reflected decisions and actions covering areas such as strategy, research and development, marketing, controlling, etc. are essential parts of the system 2 (thinking slow, following Kahneman) of an organization. But even in these areas of the organization, there are always traditional behavioral patterns and processes practiced, that have to be "rethought" when they are scrutinized more closely and consciously. The management should be very aware about that and they should specifically try to identify those "rituals" or "blind spots", question them and possibly try to change them with professional change management—they thus lift these issues from the subconscious mind of the organization into the conscious sphere. This should be done with the aim to automate the improved and more efficient processes again later on—maybe forming "hybrid intelligencies" (see Chap. 6), however monitoring them regularly and scrutinizing them where appropriate.

The areas assigned more to the system 1 of the organization with all its speed and efficiency—if the behaviors and processes are functional—are therefore closer to the subconscious mind of the organization. This includes manufacturing, call center, customer care, maintenance etc. and in particular IT operations with their infrastructural role—thus also providing the organization's connectedness to the external spheres related to the organization.

As always in business it is decisive to meet the trends in your business area. Trends are changing too. Keep that in mind when rethinking your organization and shaping the subconscious mind of your organization. Referring to the retail scenario but also applicable to other scenarios the question arises what is going to go digital and what is going to stay analogue. People will love mainly analogue experiences in future too—may it be digital immigrants or digital natives. Maybe the blending of digital and analogue user experience is in the beginning as a counter-movement or parallel development to digitalisation, for example with music "from vinyl" in addition to music-streaming or Café's with chess boards instead of playing chess with the computer (see Sax 2017). I guess this "blended approach" suits best to what can be expected from future. This will provide new opportunities for example to retailers and services close to the costumers. "Analogue is the new Bio" could be an additional emerging trend, knowing that "Digital" will stay dominant. The question is who will be able to afford analogue experiences in future. Maybe old rich people will be cared for by humans in future, the rest will be cared for by care robots. Maybe rich people can afford the sandy beach in reality whereas the less rich and poor travel this beach in virtual reality.

4.8 Complexity and Uncertainty

Organizations—facing more and more volatile political and economic environments and being confronted with new technologies and therefrom with the challenges of the digital transformation—are exposed to more and more complexity.

New "geographies" of risks emerge—the World Economic Forum is working on global risk maps—new inequalities emerge and small scale actors can have destabilizing consequences—with even cascading consequences. Organizations as part of our societies are massively affected by these rising degrees of complexity too.

A traditional remedy always was reducing complexity. For sure it is good to try to reduce complexity whenever possible especially with bureaucratic processes that do not add value concerning the organization's mission. But a lot of complexity will remain and that's getting more and more, due to the dynamics caused by the interdependencies between its parts and induced by this increasing connectedness and non-linearities.

Following Nowotny (2015) risk management was long considered one essential tool to manage complexity. Risk basically is socially contextualized being different in poor societies with people fighting for their daily living compared to western societies who feel the danger not to grow any more or even to loose prosperity. Moreover there are systemic risks typical for complex systems—barely visible risks, emitting only weak signals, interconnections not being visible etc. In many organizations risk management was established for nearly everything, with formal procedures, thus often shifting responsibility up along the hierarchies to the higher echelons and owners of the organizations, with the management levels below talking, warning and making proposals to the levels above. Risk perception, thus framing social settings as well as mood, became a crucial factor in and for the organizations and finally led to the risk paradoxon ("real" risks decline while perception of the risk increases) "Worrying about the risk is a way to cope with uncertainty" Nowotny says.

Eva Nowotny has provided a very compelling—and "widening"—view on that in her book "The cunning of uncertainty". "Complex systems cannot be left to themselves, they require a new kind of ethos arising from a sense of responsibility for being part of the whole", Nowotny says (Nowotny 2015).

An institutionalized coping with complexity is required. Prevention is usually negatively connotated whereas preparation anticipates positive and negative consequencies. Preparation as approach to cope with complexity is better than prevention. We should beware from those who want us to believe that they can master complexity (like right wing politics from a society perspective and simplifying "leadership"—managers and management guru's in the world of profit and non-for-profit organizations).

Nowotny cleary states that mechanisms must be in place that permit errors and allow performance to fluctuate, so that resources can be managed with a sufficient degree of freedom.

So what can we do in organizations to reduce complexity without simplifying too much and thus ignore warning signals and ignore the largely invisible subterranean processes as well as all the "known unknowns" and "unknown unknowns" in the background. One way to reduce the visible complexity and to cope with uncertainty are modeling and automized algorithms (Nowotny 2015). This has to be combined with human judgement—recognizing that algorithms still need to be interpreted. But where and when to evoke human judgement and where to stay with automized algorithms?

Human judgement has to form parts of an evaluation culture in an organization. Human judgement means reflecting intellectual preferences, human judgement needs to be shared, it is anchored in social communities, embedded in organizations —and it needs to be cross-validated. Complex systems cannot be left completely to themselves.

Sometimes timing is left to algorithms (for example when to place a new order in which quantity etc.). Then uncertainty seems removed but it has not disappeared— just shifted to higher level of complexity (for example: what to do with high stocks of articles and the corresponding capital engaged, when demand for this article has collapsed due to uncertainty the algorithm did not include). In general, complexity of social life increases with number and density of interaction. What we can learn from biological systems and evolution is: the design must be robust and enable resilient responses.

That is where the metaphoric view on organizations and this model of the subconscious mind of organizations can be helpful: Find the adequate mix of reducing complexity by finding the right mode of governance between decentralization and centralization, between automatic algorithm and human judgement, between "thinking fast" and "thinking slow", to provide the necessary flexibility, self- regulation and thus keep uncertainty at bay.

Good governance is not only provided by a strict integrated management system regulating everything, it also must allow a little bit of chaos or muddling through, improvisation and self-organized response. Good governance has to provide stability but also has to respond to instability, it must give room for pragmatism, incrementalism and piecemeal approach in an organization, it has to be adaptive—and it has to be incorporated in an organizational model as presented in this chapter.

4.9 The Consideration of the Model as a Whole

As a round-up of the model description, now an attempt to a further holistic perspective on the subconscious mind of organizations and its internal cascading:

- Every employee has—as part of their personal subconscious—their "subconscious with professional respect" and is integrated in a group, in an organizational unit, in the infrastructure of the organization and the environment of the organization.
- Each organizational unit and group has its own "organizational subconscious mind", which is characterized by its specific task, its infrastructure and its connectedness to the organization and its environment.
- The organization as a whole has an "organizational subconscious mind" including its infrastructure and its connectedness to its environment.

Even when the top management of the organization is faced mainly with this holistic view of the subconscious mind of the organization and will try to use this as best as possible, there is one essential aspect to be considered by top management: it must also be aware of the specific subconscious mind of the organizational units and employees to a certain extent in order to align the entire company as good as possible and to condition its dealing with internal contradictions and trade-offs in decision-making at the relevant level correctly—it has to provide adaptiveness.

One might think: That always was the objective of organizational work. This is of course correct. But the new technologies and the understanding of the subconscious mind of the organization as a combination of the subconscious mind of the respective organizational units and the "professionally relevant part of the subconscious" of employees lead to new, more holistic approaches and design options.

How this subconscious mind and its collaboration with the conscious mind of an organization can be designed best and how—in the design of the traditional and necessary corporate and organizational functions—this can be influenced in the right direction, is described in the next chapter of this book, which is dedicated to the design possibilities and the shaping of the subconscious mind of organizations and its interfaces to its conscious mind. This again is illustrated with stories and forward-looking scenarios from a fictitious industrial company, a fictitious retail company and a fictitious organization delivering health care services (a hospital group). Furthermore guiding principles (GP) for shaping the subconscious mind of organizations are provided inj the following chapter.

Literature

Forrester Research (2013) US Online retail forecast 2012 to 2017. Accessed online 3 Mar 2013
Nowotny E (2015) The cunning of uncertainty. Wiley
Sax D (2017) Die Rache des Analogen. Residenz-Verlag

Chapter 5
How to Form and Shape an Organization and Its Subconscious Mind

Abstract The theory of complex systems is based inter alia on systems theory and chaos theory and their findings concerning the emergence. "The whole is more than the sum of its parts" can serve as an overall metaphor for that. This chapter about the formation of an organization and its conscious mind as well as its subconscious mind follows the analogy to the processes of the human brain as source of creativity and innovation: the perception, the cognition, the knowledge and the processes of decision-making and interaction. Innovation in itself and the prerequisites for successful innovation are the basis for guiding design principles presented in this chapter. They are then expanded to include aspects of organizational culture thus leading to practical guiding principles (GP) for the design of the organizational structure, the process structure and the infrastructure as parts of the subconscious mind of organizations as well as the embedding of the organization in the respective business area and society. The development of an integrated management system as framework for quick and flexible action is recommended. The extended perception, the extended—implicit and explicit—knowledge and the consciously and disciplined, semi-automatic or automatically derived findings, conclusions and decisions shape the subconscious mind of organizations and the interface to their conscious mind. The "new thinking" of an organization in the light of new technological options such as Artificial Intelligence, automated decision making etc. as well as coping with the uncertainties and upheavals in the environment of the organizations require time, effort and discipline. It is therefore one of the key management tasks to face these developments and challenges actively and not take the "easy way" to leave this to the daily business or simply "let it happen". The sustainability and long-term existence of an organization is being massively affected by the managerial approach chosen. Finally humans hopefully will still be—in terms of governance—"in the driver's seat". If we do it right, future organizations will not constitute as "Artificial Intelligences"—as part of Ray Kurzweils "singularity"—, but they will develop and evolve in a kind of "hybrid intelligence". Concerning the people and their relationship to the structures, organizations and new technologies we should basically follow the principle: "People do not serve structures but the organizations, their infrastructures and their systems serve the people".

Keywords Complexity · Artificial Intelligence · Innovation · Organizational sustainability · New thinking · Integrated management system · Decision making

The underlying thesis of this chapter and actually the main idea of this book is, that the subconscious mind of organizations—as defined and described in the previous chapters—can be formed and shaped.

Let us first of all try to refute this thesis. That would mean that the system behavior of the subconscious mind would develop from the interaction between the individual elements and the subconscious mind's interfaces to and its interactions with the conscious and the awareness—the subconscious mind would thus be simply "emergent". Thus it would not be necessary to widen the view and it would be ok to stay restricted to the obvious and visible design of the organization. It would not be necessary to rethink our organizations more fundamentally—they would then be already developing in the right direction. It would be relatively easy to handle the new technologies and digital transformation—let the subconscious mind of the organization emerge and let us finally have a look what organizational behavior emerged.

Let us look at the idea of emergence in this context a little bit closer: Emergence is described as spontaneous formation of new properties or structures of a system due to the interaction of its elements (see Stephan 1999a, b). This is a valuable complementary observation when we look at the design of the subconscious mind of organizations and on the other hand it is a clear indication not to think that everything can be designed. The subconscious mind of a person also is not designable—but it can be influenced by learning, living conditions, social environment, rules etc. Isn't this to an even greater extent applicable to the subconscious mind of organizations? The human subconscious is located in the brain, it is technologically inaccessible, only to a small extent it allows us to take influence biochemically—until now. Whereas the infrastructures of organizations and their design are indeed accessible by humans in the way of purpose-oriented design. So we can design our organizations but we also can shape the subconscious mind of our organizations—and not let that simply happen and "emerge".

Developments in systems theory and chaos theory show (see Gleick 1987) that phenomena such as self-organization and their conditions of formation are quite accessible to systematic and objectively reasonable explanations. Self-organization may well be referred to as a phenomenon close to emergence. The context and conditions of emergent systems are largely in line with the characteristics of self-organizing systems. Feedback processes on the basis of self-reference or circular causality play an important role in these systems. Evolutionary biologist Ernst Mayr defines: "Emergence in systems is the appearance of features on higher levels of organization, which would not be predicted simply on the basis of known lower level components." (Gibb 2011).

Reasons for this are:

- The system is already so complex, that it is not examinable or not able to be simulated without reduction.
- Between system elements new compounds, effective relationships and processes arise that have not been implemented (pre-planned).
- The couplings or effective relationships between all elements are changing with the integration of a new element.

System theorists believe emergence is a distinguishing characteristic of hierarchical systems. Such systems have features at the macro level that are not present on the simpler organizational level, the micro level. They are formed by synergistic interactions between the elements at the micro level. The theory of complex systems is based on insights about the emergence from systems and chaos theory. A popular saying summarizes this very simplistically: "The whole is more than the sum of its parts."

Human contents of thought (ideas, concepts) have characteristics of emergence concerning neurological processes and mental acts, from which they arise. Likewise emergence effects can be seen in the communication of content of thought, because the characteristics of information can not be derived linearly from the underlying grammatical structures (letter, word, syntax). A very striking description of this is another popular saying: "Truth is not what I say, but what the other understands."

In connection with the new media—such as the Internet—we can also speak of emergence. The Internet allows new effects to arise, which can be described as emergent. By further linking, these effects are amplified-the network effect. Examples for that are "Internet art", "Smart Mobs", "Shitstorms", online games, Internet forums, wikis and grid computing as well as platforms like Google, Facebook, Amazon etc.

We also often find analogy considerations between the behavior of swarms in nature and the behavior of gatherings of people with similar interests: Size, form, shape, direction, speed and wave motions in swarms are emergent relative to the individual, for example with fish or birds. Their movements and changes of direction partly seem to run swifter than the isolated reactivity of the individual fish or bird would permit (Gibb 2011).

The above shows that both design and emergence will have their importance in the consideration of the subconscious mind of organizations and it will have to be an essential design feature—thinking our organizations new—to watch closely the developments resulting from emergence and to strengthen them or mitigate them or counteract, wherever that makes sense. Emergence is a great opportunity for innovation. In addition to the effect of emergence there is also some configurability given for the subconscious mind of organizations—especially in the field of infrastructures.

Basically, it is important to ask and answer the following questions if we want to distill systematic and practicable guiding principles for shaping, designing and "rethinking" of organizations from analogy considerations between the subconscious of the human individual and the subconscious mind of organizations:

- What do we need to consider in the design of systems and organizations?
- What are the opportunities and threats to be considered?
- What are innovative organizations and how do they behave?

The guiding principles are summarized in the paragraphs marked with GP. At some points they are exemplified by the fictional scenarios of the Hospital Group, the special steel company and the retail company, as we did in order to illustrate the model of an organization (Chap. 4) and the therein described operation of the subconscious mind and the conscious mind of organizations.

Let's start with a guiding principle (GP) carved out from the above dialectical consideration of emergence and design related to system behavior:

GP1 Analyze the systemic behavior of your organization!
Observe the systemic behavior of organizations and recognize potential emergent developments resulting from the (complex) interaction between the individual components and subsystems. Recognize the potentials for innovation and risk.

Example hospital group—*e-mail visits as organizational innovation emerge from changing conditions*
Robert as a specialist of the department of internal medicine at a medium-sized hospital gets more and more emails from patients to his professional email address. They want to share with him, how they are doing after treatment in the hospital and sometimes they also ask him questions. 75- to 80-year-old are patients who are already familiar with the Internet and with e-mail communication frequently do it as well as younger patients. Sometimes they share observations, which they had not communicated in the particular environment of a hospital and the mostly associated stress situations. For Robert these observations were quite relevant for the management of the patient's disease. Now that they had electronic access to their health record, suddenly this phenomenon has emerged. The first time Robert had politely but firmly replied that they should discuss it with their general practitioner, who would then maybe again assign them to the hospital if necessary. Some patients then showed up again in the outpatient department, but that was useless in most cases and only was wasted time.

From Robert's perspective it would have been better to offer some kind of "email visit". But that of course was not paid by anyone given the current remuneration principles. On the other hand the outpatient clinic visit was paid at a flat rate, bringing not even significant revenue for the hospital and being far from covering the real costs for these visits. After this had happened to other doctors too and some patients even had written emails to his boss, the chief physician Sandra, she had discussed it at the department meeting and, subsequently, with the hospital management. Even Paul, the CEO of the hospital group, had been confronted with this important and fundamental issue.

Soon a strategy for dealing with this new phenomenon was found: The email visit was made possible and the senior physicians were also given a time quota for it. And there were defined rules: The patient can communicate only with the supervising or discharging senior physician and structural support was offered too, such as questionnaires for patients etc. This had the additional advantage that doctors could now track the quality and the success of their treatment better and they also could save numerous follow-up visits in the outpatient clinic—and probably at other physicians too. These would have costed much more time than a paid email visit. The number of patient-contacts at the outpatient clinic already had noticeably reduced with this new means of communication.

After a few months of experience with these well managed email visits someone said in the department meeting: "It is interesting that of various elements such as increasingly "email-literate" patients, the introduction of an electronic health record, and the announcement of the company's email addresses of doctors suddenly something new and something excellent emerges—if we actively manage it".

Constituting the subconscious mind and deriving the design of the subconscious mind of an organization and especially its interfaces to the conscious mind based on the model in Chap. 4 and the explanatory scenarios, it seems obvious at a first glance to work along topics like research and development, marketing, sales, production, finance, controlling etc. as these are now the most important functions of an organization or a company.

Let us, however, view organizations in a new and different way and design organizations and their conscious mind as well as their subconscious mind in accordance with the processes of the human brain as a source of creativity and innovation. So let us choose the following approach:

- Let us create "re-imagined organizations" along the everyday—often partly unconscious—processes of perception, recognition (and its use of knowledge) as well as the processes of taking decision and action (Sect. 5.1).
- Subsequently, let us discuss some fundamental aspects of innovation (Sect. 5.2.1) and elaborate on the innovation of systems and infrastructures (Sect. 5.2) as well as the innovation of an organization (Sect. 5.2.2)—as far as the previously derived design aspects of organizations and their conscious and subconscious mind are associated.

- Based on that the impact on culture and consciousness of an organization will be described as well as the relevant associated design options (Sect. 5.3.2).
- Next, the aspects of the organizational processes (Sect. 5.3.3) and the organizational structure (Sect. 5.3.4) are deepened.
- The synopsis of all these infrastructural aspects and innovative organizational considerations require a frame for the holistic and systematic management of an organization. So a suitable form of an integrated management system will be outlined (Sect. 5.3.5).
- Here additional relevant aspects are discussed, the role of trust, the management of collaborations as well as a culture of excellence (Sect. 5.3.6) and the topic of resilience (Sect. 5.3.7) of organizations.
- Finally and to round up,—after respective considerations about managing complexity and embracing uncertainty—organizations are viewed as a special form of ("artificial") intelligence—as an entity composed of diverse elements with humans, Artificial Intelligences and hybrid intelligences among them.

5.1   From Perception to Decision

We therefore—as established above—try to figure out "re-imagined organizations" along the everyday, often partly unconscious processes of perception, recognition (including its use of knowledge) and the processes of taking decision and action.

5.1.1   Extended Perception

The technical developments and possibilities of virtual reality and augmented reality were already treated in Sect. 3.2.6. The possibilities of perception of humans —and thus also the abilities of organization for perception—are expanding.

After the widespread proliferation of smart phones a wave of new, wearable devices ("wearables") are available now as the smart watches, transparent "smart" contact lenses serving as monitors, EEG-based headsets as Brain-Computer-Interface (BCI) voice computers in the ear etc. These augmented realities penetrate into our perception in various ways. In parallel, abilities of perception of the computers themselves are expanding enormously. It is no longer just about what humans will see in the future, but also about what computers will "perceive". The chip giant Intel talks about the concept of "perceptual computing"—processing perception in new ways and realtime. The computers and tablets will be able to perceive the depth of space. In the new camera modules from Intel an infrared source is integrated. The reflections let the computer perceive where corners, edges and faces are. The computers make a 3D image of their surroundings. This is what will significantly enable the computer to be controlled through gestures (see Schmitt 2014). These scanners of the environment which partly already can be found in middle-class cars, coupled with navigation equipment and new projection technologies (projection of display on the windscreen), will enhance the driving assistance systems significantly and help with keeping a course, showing deviations from the track, etc. and lead to autonomously driving cars.

What is once digitized, can be linked with other digital data, and compared with the new real objects. They will help in answering the question: What does it mean? The next challenge is to search the real world. Pioneers in these technologies forecast a full 3D objects Wikipedia. Thus "the world will become machine-readable". The machine learning in the context of Artificial Intelligence—the ability of the computer to detect patterns and derive regularities and then learn from own experiences is within reach.

So are many new ways to extend and expand the performance of organizations in the making. This increasing technical capabilities of sensors and their networking supplies us with (partly realtime) data in a volume, a quality and a diversity that make us ask the question: How should we deal with it? Driven by algorithms for pattern recognition and visualization of large data sets new perspectives can be gained about the reality.

Another aspect of the "extended" perception of organizations are the social networks. The connection of the customer can be designed in new ways—supplemented with the new techniques of augmented reality. On the other hand one or another problem can be detected early by observing what appears about the respective organization on the Internet. Thus an impending "Shit Storm" can be averted by timely proactive information or the like. Many offered tools for social media analysis already can provide valuable assistance here.

In the interactive dialogue with the data, the perception of an event and of the environment by the organization can be further sharpened and can lead to new insights. More about that later in the chapter on knowledge (Sect. 5.1.2).

GP2 Explore the potential extended perception of your organization!
Search for opportunities offered by the new technological possibilities for the respective organization in order to enhance the perception of what is going on and how the organization itself is perceived. Anchor these opportunities within the organization and link them appropriately in such a way that this expanded and extended perception—where relevant—enables corresponding findings and actions.

Example hospital group—*e-mail visit discussed in social media accelerate solution*
The organizational unit (OU) Corporate Communications of the hospital group had already discovered on the Internet and on Facebook some negative reactions that the Department of Internal Medicine had not answered emails of patients with respect to their treatment. Paul, the CEO was informed and

alerted. The "advanced perception" by observing the social media had certainly facilitated an accelerated finding of a solution with a regular handling of patient emails concerning medical issues (as described above) in a secure environment concerning data privacy. The quick reaction had possibly prevented a controversy and bad press on the internet.

Example industrial company—*develop a new "feeling" for complex dynamics of the rolling mill*
Frank, in charge of the maintenance of the rolling mill, motivated the mill suppliers to install him a monitor. The supplier had installed sensors to record vibrations and linked them in order to monitor developments in the field and in the long run to provide better prediction for prevention of machine failures. The recorded data were graphically presented in the installation drawing related directly to the aggregates. Data for a period of about a month were made available and he could graphically compare several measuring points. Whenever he found time he "played" with the new features on his monitor. He felt that he could now get better acquainted with his rolling mill and that he could develop more sense and "feeling" for his rolling mill and its complex interdependencies and dynamics than ever before. Frank was now able to clearly see the load profiles as well as downtimes of individual aggregates. Frank realized that virtually all operating data that have been provided by various system parts, would be derivable and thus predictable from these new sensor data. He decided that he soon would discuss these new opportunities, which were made possible with the "extended perception" by the new sensor technology and the algorithms with the rolling mill manager Ralph and the project leader Gabriel. These sensors that were actually implemented for maintenance purposes indirectly provided operating data that were very interesting—the rolling mill operation parameters could be derived, which were crucial for their performance and their quality and not supposed to be transferred somewhere else.

As there are offered new ways to perceive complex situations and dynamics by good visualisation based on new data provided by new sources (like sensors on the bearings of the rolling mill, opinions and comments about my organization in social media or vital parameters patients take at home like blood pressure, glucose level or weight), decisions will improve when based on more and new relevant information. The approach of the individuals who have to decide or teams who discuss and then decide will change with such new ways to perceive—the subconscious mind of the organization is thus changing, hopefully improving.

5.1.2   Knowledge, Memory and Recollection

For the understanding of knowledge it is appropriate—especially when it comes to the area of information and communication technologies, as in this book—to clearly define the concept of data and information and relate them to each other: data are generated for example from results of measurements or series of observations. But these numbers are worthless per se. For example, the measured number 38 has to be brought in a context. Only an associated syntax will turn it into an information, e.g. Mrs. Hofer has 38 °C body temperature. We can speak of knowledge if information and conclusions are generated from data and information and if necessary actions are derived—thus the data become "actionable" (see Sect. 3.2.12 about knowledge management). To become actionable, relevance has to be attached to the information which, in turn, results from the context of the situation and the people involved. An example: 38° body temperature has a different meaning to the emergency management of an unconscious patient than for the General Practitioner, confronted with a patient who complains about weeks of fatigue and reports that he had elevated temperature (38°) for weeks, but that he had no cold. With their knowledge, that they retrieve in their specific situations, the emergency physicians and general practitioners come to different conclusions in each case and probably take different action. So far—so trivial.

Knowledge occurs in two forms: explicit knowledge, which is verbalized and displayed, so that it can be stored or archived, thus transferable. A very important different form of knowledge is the implicit knowledge—also called "tacit knowledge". This "silent" knowledge includes not only cognitive structures, but also mental processes that are difficult or impossible to verbalize. It is about mastery and experimental knowledge (following Neuweg 1999). This is best expressed in the popular saying: "We know more than we can say".

Our brain operates spontaneously, automatically and intuitively—with all its side effects, as they are depicted by Kahneman in his behavioral psychology. "The knowledge is in our ability."

Humans have this implicit knowledge, but also teams have it. In teams each individual for his part, but also overlapping all together know "how it goes" (see Springer Gabler Verlag (Hrsg) 2011). This cannot be detailed in process descriptions and it is only effective when—in spontaneously functioning experienced teams—it is "intuitively" available and usable.

The new technologies enable this by the methods of pattern recognition—based on the rapidly growing storage and processing capacity—thus enabling new ways to store knowledge and access it (in a kind of "extended" memory and remembrance). The ability to have access at a specific situation to patterns of action or behavioral patterns of the past and to be able to make comparisons with the actual situation and its patterns, to determine the relevance etc. are new opportunities enabled by Big Data technologies and their algorithms. Since the results of past action patterns can also be identified, in the situation of knowledge access and decision-making also forecasts from the extended memory can be derived.

Part of what is available as an implicit, procedural knowledge—but only a part of it—can be made available and accessible this way for a broader group of employees and is therefore part of the subconscious mind of an organization.

GP3 Explore behaviorial patterns of your organization from Big Data!
Look for ways to use in addition to the stored factual knowledge the new technologies available and the "tacit knowledge" in the form of patterns of action and behavioral patterns out of available Big Data in order to make relevant comparisons to current situations of action to make it available for decision-making. Motivate and enable your staff to assess the relevance critically.

Example hospital group—*detect impending sepsis earlier than an experienced physiscian with predictive analysis*
Sandra, the chief physician of the Department of Internal Medicine, has heard, through ongoing contacts with the Department of Anesthesiology and Intensive Care, of already internationally piloted systems where it was possible to detect the warning signs of impending sepsis earlier—about two hours before this was possible for a very experienced doctor. This was enabled by the purely statistical comparison of numerous similar cases in the past with the continuously captured data from an actual ICU patient and his patterns of change in his vital signs ("my patient—like my patient"). Countermeasures taken two hours earlier could possibly avert serious harm from the patient—and help also save resources incidentally. The enhanced perception associated with the use of "expanded memory" obviously made new dimensions of "tacit knowledge" accessible and usable. For her doctors these were also learning experiences that expanded their horizon and their sense for critical medical situations. This sense would also improve the intuitive action and the possibilities of intuitive plausibility checks of facts and decision options. Sandra wanted to bring to discussion these project ideas and the possible use of it in her department at the next department meeting …

Example industrial company—*hidden knowledge in vibration patterns*
Gabriel, the project manager of the innovative maintenance project, became aware after numerous discussions with the engineers of the mill manufacturer which hidden knowledge lay in the—at first glance seemingly worthless—vibration data of the bearings and rolling stands in advance of past disturbances. But you would need the right algorithms for pattern recognition. That

for him was a fascinating new world of "extended perception" by sensors on the one hand and the knowledge from data on the other hand. As a young engineer he had—with this project—a singular learning opportunity for his further career.

Knowledge and the memory of the organization as well as the tools to use it and remember it in a context-sensitive way are augmented and thus—well implemented and organized—improve the organization's subconscious mind when used in decision making of individuals and in teams—hopefully on as many hierarchical levels of the organization as possible.

5.1.3   Cognition, Decision and Action

If we want to continue the "cognitive process of organizations" in analogy to the cognitive process of humans—after discussing the perception on the one hand and the aspects of knowledge, and of memory on the other hand—now the aspects of recognition, knowledge discovery, decision making and setting of actions have to be considered and to be discussed from the perspective of "thinking organizations new":

A saying, that is attributed to Immanuel Kant reads as follows: "The need to decide exceeds the ability to detect". The analysis and intuition, as well as environmental conditions such as organizational and social integration are to be considered exactly, given the uncertainty inherent in decisions, and the fuzziness of the initial situation and the incompleteness of decision basis at the point of decision. This saying of Kant applies on the decisions of individuals but is applicable as well to group decisions and decisions of organizations.

Reinhard K. Sprenger, an eminent thinker and lecturer on topics such as motivation, self-responsibility and trust, formulated the subjective component in decision-making as follows: "I decided because of certain reasons, of which to me some are known, some are unknown, but in any case the decision has me as a person as a prerequisite. "I"—this is not random, but it is a man with his—only his —history." Sprenger also mentions the comparison with the iceberg: only 10% of the volume is visible—that is our conscious mental activity—90% are under water —these are the unconscious mental processes "I am not forced, I do not act randomly, I experience myself as the author of the decision" (Sprenger 2013).

This personal view of decision making needs to be considered in any case—even if decisions are taken by organizations—except automated decision algorithms. But as soon as the human factor is actively involved—as in decision support systems— this "I" is taking effect.

But this "I"—to a certain extent—is influenced in its integration into the organization, it can be conditioned, it can be shaped by a deliberately designed subconscious mind of the organization. Vision, mission, strategic and corporate goals, corporate culture, organizational development, staff development, training etc. are elements to achieve this.

Develop and live a jointly supported vision

Starting from a clear picture of the reality an image of the desired reality—the vision—can be developed in a discussion process. Then strategies can be developed in the form of strategic directions of impact and the related strategic objectives for achieving that vision. A vision shared by the organization should be powerful and vivid as a hologram. Some describe this as "Big Picture". The individuals and organizational units should not only behave in accordance with these objectives in terms of "compliance", but should also develop a clear commitment for the vision, the strategies and the objectives. In the highest form that creates a common purpose, which is pursued in partnership with each other.

It is of particular importance in the decision process and before setting actions to practice a certain analytical discipline of thinking—and to train this. The correct thought patterns—the way how findings are derived from the facts and how these findings and insights are developed towards conclusions and recommendations and how these finally result in the decisions taken and actions set—can be a great help.

Establish a discipline of thinking and corresponding mental models

Systems thinking is an essential skill that should be conveyed to—above all—decision-makers but also to as many employees as possible. Basic Concepts of structured thinking, how they are used by numerous business consultants usually in a very disciplined manner, are for example the clear distinction between facts, findings, conclusions and recommendations. The way of the facts to the recommendations has to be presented clearly and comprehensively and underlying assumptions should be clearly stated—a kind of "logic diagram" should be set up. Since most are complex issues and problems, it is useful—starting with a clear picture of reality and of the relevant environment—to first formulate working hypotheses. The next step is to try—in the selection of the relevant facts and findings, conclusions and recommendations—to substantiate or disprove these working hypotheses through appropriate questions and therefrom selected facts and findings. The result is a coherent—albeit requiring some effort—but structured and accountable process of reflection from relevant facts to relevant findings and then to conclusions, recommendations and finally decisions—embedded in working hypotheses. Prejudices, backgrounds and similar traps that Kahneman (2011) in "Thinking fast and slow" described are thus largely eliminated. Such a complicated process of reflection is clearly assigned to the system 2—thinking slow.

Managers and staff of organizations should master such decision methods. If they are well practiced and internalized, they can also work in shorter and less formalistic processes for less complex issues and so ultimately lead to an intuitively guided, better quality of thinking and to coherent arguments. The best exercise is when you regularly—with decision cases selected by chance—scrutinize decision proposals and decisions using such thought patterns and let the humans in charge have them explained and discussed.

To develop such qualities in cognition processes and decision-making processes and to establish them, is a difficult and longterm task to be accomplished. The leaders, managers, team leaders etc. must become champions in this discipline of decision-making and they should ideally lead the learning processes in their teams accordingly. This does not necessarily require expensive seminars with renowned speakers. The learning process in these disciplines of thinking and of mental models can be practiced continuously in everyday decision-making in the department meetings and everyday conversations by understanding and using these as learning situations, scrutinizing and discussing the analytical derivation of presented decision proposals. Executives—but also very capable members of the staff have indeed to take the role of "learning coaches" in these situations. It is to establish a culture of "Team Learning". Even experienced staff members at the end of their careers, for example, in partial retirement, etc. can bring in their experience as facilitators and coaches for the benefit of their organization.

Develop a culture of "Team Learning" and reasoning

The learning of teams and within teams—be it in organizational units or temporary working groups according to the model of an organization (Chap. 4) or in virtual teams—is crucial for the conditioning and training of the conscious mind and thus the system 2 (thinking slow) of an organization. But it conditions by means of the training and the training effect also the system 1 (thinking fast) of the employees by framing and shaping individuals and thus possibly improving their intuitive actions. Thus decisions proposed from the subconscious mind of the organization and its System 1 (thinking fast) become better and System 2 (thinking slow) of the organization will be called at the right decision points only. Thus decisions get moved to the right level of awareness in the organization. The learning processes can be varied —basic learning projects, project reviews, trainings, seminars, workshops, technical reviews, regular reviews like Morbidity and Mortality meetings in medical teams, etc. But they should be consistent in the organization—supported by a jointly supported vision and thought in similar mental models and lines of argumentation.

The above rules should be applied to semi-automated decision-making processes as well, in which people are supported by decision proposals from decision support systems (DSS), because these DSS are often seen as a relief from individual responsibility and the individuals and teams may "blindly" follow the decision proposals. Decision support systems should therefore be designed in a way that the backgrounds of their proposed decisions are to be made reasonably understandable —at least roughly. This reasoning should always be offered at least to evoke the system 2 (thinking slow) mode at the level of the individual person, whereas it maybe stays in the system 1 (thinking fast) mode at the level of the organizational unit—in the view of the organizational hierarchy above.

We could now argue that this reasoning is useless—facing Big Data and sophisticated algorithms. Samuel Arbesman, a complexity scientist and writer, published his book "Overcomplicated: Technology at the Limits of Comprehension" (Arbesman 2016). It's a well-developed guide for dealing with technologies that elude our full understanding. In his book, Arbesman writes we're entering the entanglement age, in which we are building systems that can't be grasped in their totality or held in the mind of a single person. In the case of driverless cars, machine learning systems building their own algorithms to teach themselves etc. and in all the processes that become too complex to reverse engineer.

Though we won't be able to "reengineer" the decision making process and the reasoning behind decision proposals in detail and completely, we sharpen our gut-feeling of what is going on when continuously trying to find the reasons for the proposed decisions and scrutinize them at least at some of the cases.

A special form of decisions are purely automated decisions. This purely automated decision-making, which is indeed increasingly implemented, requires good, random-led review processes by individuals and teams. Therefore it is necessary to document the reasoning from the algorithm correspondingly traceable to challenge the decisions taken in the review on an appropriate basis and to allow a learning process and the improvement of algorithms and their implementation.

GP4 Deploy a vision!
Develop and live a jointly supported clear vision as a framework and background for the countless conscious, intuitive and subconscious decision processes in the organization.

Example hospital group—*a strong vision brings benefit beyond your own organization—for the patient*
Paul, the CEO of the hospital group had worked out the strategic direction of improving the patient-centered collaboration of hospitals in his group in teams together with the admitting physicians and other established structures of nearly all relevant healthcare providers and had also agreed that with the owner of his group. The measures to improve the discharge management were already applied successfully at the internal department, using the millions of case histories of the group from the last 15 years as a basis. These measures would have probably not been financially viable and feasible for his group only without this strategic commitment and active collaboration of stakeholders, decision makers and other partners in the provision of healthcare, because the benefits of these measures would only indirectly benefit his hospital group, but mostly benefit the patients and the healthcare system as a whole.

GP5 Enforce individual and organizational learning!
Develop a conscious approach to human, semi-automated (decision support systems) and automated decisions in the organization and observe and deliberately design and shape the individual and organizational learning processes. Attempt to determine the actual results achieved consistently in order to feed back the effectiveness of the improved decision-making processes to all acting and involved people and organizations and thus establish a systematic learning process.

Example hospital group—*readmission rates improved through algorithms and team learning*
Sandra, the chief physician of the internal department was aware of how dangerous it would be if suddenly a large number of automatic decisions were taken. Therefore, this project to support decision making at the discharge management was ideal as an entry point into this new world of decision making. It was only the probabilities of the possible causes of the forecasted readmission that were proposed to the hospital staff. The decision of the measures accompanying the discharge of a patient was still matter of their staff. The discussion of surprising results with this new way of decision support regularly took place in the Department Conference. So a learning process for the entire team was started. Already the first results have been discussed in terms of the actual readmissions after the introduction of the decision support system and the improved readmission rates. They were promising, but the observation period was obviously too short. Even doctors from other departments took part once a week at the team meeting, as the hospital management wanted to promote the wider use of this system.

It was common opinion that decision support systems should be used to improve quality and not to accelerate work. If there was not enough time granted for reasoning and evaluation of decision proposals—both on individual level and in team-learning sessions—the knowledge and the skills of physicians would erode and quality would be in danger in the long run.

GP6 Implement a discipline of thinking/mental models!
Establish a discipline of thinking and corresponding mental models for the derivation of decisions and pay attention to the transparency in terms of reasoning concerning decision proposals and automated decisions taken.

Example industrial company—*improved decision making through transparent assumptions and reasoning*
Gabriel, the young, curious and ambitious project manager of the special steel producer had learned to structure the decision making process at a seminar of the consulting firm in charge of accompanying his company with the question of future maintenance strategy. He had firmly planned to surprise his superiors with a clearly structured decision report, where the facts, the findings, the conclusions and the recommendations, as well as the underlying assumptions and their reasoning were clearly worked out. He had never actually experienced such structured argumentation in his company so far. Actually more discipline in thinking would be good for the company as a whole, he thought for himself. Maybe he would contribute to general improvement in this area of decision making culture with his work.

GP7 Continously improve decision making!
Develop a culture of "Team Learning" and continuous improvement of human, semi-automated and automated decision-making processes as well as the underlying systems and infrastructure.

Example hospital group—*combining evidence based medicine with team learning*
Paul, the CEO of the Hospital Group, had just held a conference with the medical directors of his hospitals in which the use of evidence-based and guideline-based medicine was the main topic. It was not only about quality but also about forensic aspects and risk management issues. In a heated debate, as often before, the "art of medicine" and the necessary individuality of care were heavily discussed. "Evidence-based" had been affirmed by the vast majority of the participants. Concerning "guidelines based" there was a much more skeptical approach in the group, since the taking of responsibility

by doctors was in question if regularly recommendations—according to guidelines—for decisions to be taken were proposed. Finally, they had agreed to approach this topic carefully with pilot projects and to set up a longterm organizational learning process in the department's teams as well as in interdisciplinary teams and also at the level of medical directors and senior physicians. With nursing it was a bit easier. The colleagues were convinced to be able to learn from the area of nursing who already had some expertise in Evidence Based Nursing (EBN).

Conclusion
The extended and expanded perception, the advanced access to more and more specific knowledge (implicitly or explicitly) and the resulting (consciously and disciplined, semi-automatically or automatically derived) findings, conclusions and finally the decisions taken are influenced by the subconscious mind of organizations and the interface to their conscious mind. They can be actively shaped. The possibilities for design of the subconscious mind of organizations are manifold, but they need time, effort and discipline. It is therefore one of the major management tasks to face this development and the challenges consciously and not to easily leave this to the daily business or simply "let it happen". The sustainability and longterm existence of the organization will be affected massively.

In addition to the above described ways of decision-making the issue of innovation is—in the light of new technologies—of particular importance for the viability and the success of an organization and its sustainable and longterm existence. But innovation is always a question of the attitude of the humans acting and their access to the infrastructure. Let us therefore consider the relationship of innovation with the subconscious mind of organizations and their reciprocal influence and interactions in the next sections.

5.2   Innovation

Innovation is one of the most important parameters in the development of organizations and ensures their sustainable existence. What conditions do favor innovation in organizations? What can be learned from the history of innovation for current and future organizations? How do the innovations—in particular those associated with the new technologies in the information and communication technology (ICT) area—change the organizations? What effect do they have on the subconscious mind of the organizations?

Let us attempt to answer the following questions by considering some fundamental aspects of the innovation itself first and then make some specific considerations:

- How to innovate by designing the systems and infrastructure within the meaning of the model of the subconscious mind of organizations
- How to influence and shape the "intuitive skills" of the individuals as well as of teams in terms of innovation
- How to proliferate product innovation.

Then we will try "to rethink organizations"

- In the sense of innovation in culture and awareness of the organization
- In the sense of innovation through process organization
- In the sense of innovation through organizational structures
- In the sense of the tool of an "integrated management system" as a framework for an organization and its management.

and thus embrace the innovative organizational design of the "subconscious mind of organizations" and its interface to their "conscious minds".

5.2.1 Innovation per se

Johnson (2010) has worked on innovation in itself and the history of innovation in his book "Where good ideas come from" in an impressive manner. The following ideas are based on it.

He has looked for guidance in the patterns of nature (especially the patterns of coral reefs) and he has compared it with the innovation and creativity patterns of large cities and the "network" (Internet). He worked on the characteristics of "self-similarity" (as it also appears in complexity theory and chaos theory) arguing along the line "if successful in nature—why not in the organization". He deduces that self-similarity gives orientation and that it requires standards in infrastructure and methods—standards with a certain variability (in order to allow innovation)—, similar to the successful patterns in evolution.

Following Johnson, megacities and the Internet are proven spaces of innovation. With our innovative ideas we shape the environment in which we are living, but the environment does the same with us. Environments with an innovation level above average have certain characteristics and patterns in common. He has developed a unified theory that describes the common features of these innovative spaces. He thinks that, if we approach a problem interdisciplinarily and if we are looking for self-similar fractals, we can create new knowledge and find similar patterns of creativity that recur at various levels.

I think if we want to claim "to rethink organizations", it's worth to try to find relevant similarities from these patterns of innovation—but also to recognize differences.

Johnson has observed nature by comparing the patterns of innovation at all levels (global evolution—ecosystems—species—brain-cell) and observing culture on different levels (ideas—environments—organizations—cities—information networks).

Johnson has then derived the following innovation patterns. I have processed that with relevance to the subject of this book and supplemented it with proposals for the innovative design of the subconscious mind of organizations. These guiding principles are of course especially relevant for the design for organizations that require particularly high innovation, but they also provide valuable hints for other organizations.

The Adjacent Possible

The history of the human species and its cultures is a series of experiments to explore the adjacent. The trick is, to thereby stretch the possibilities currently given. Often it is enough to change the physical working environment, to cultivate the social network, to manage your own information better, etc.

> **GP8 Explore the adjacent!**
> Try to explore the adjacent possible: There are systems and organizations that succeed better to explore new realms of possibilities and opportunities. Look for organizational improvements focusing on and following this idea—the adjacent possible. It is not always the far reaching idea that matters.

Liquid networks

The network must be adaptable. It must be able to interconnect itself anew, to explore the boundaries of the adjacent possible. Innovation requires a "liquid network" of high density, for innovative systems have a tendency to chaos. A "fluid environment" provides sufficient stability, so that innovations do not dissolve too early but can grow. This may enable "Flow". Flow is a feeling like to float in a stream which is carrying us in a certain direction, but always giving us new and surprising impulses with its whirls and waves (following Csikszentmihalyi 2010).

> **GP9 Let ideas grow in a "liquid" organization!**
> Encourage and allow "liquid networks"! In order to make the spirit more innovative, we must ensure an environment as networked as possible and working conditions that allow to focus. "Flow" is created when the mind is totally focused and at maximum productivity.

The Slow Hunch

It often begins with a first thought. The gut feeling or "emotional brain" that—in its flash estimate of the situation—overrides the much slower working logical thinking and in the end might even be right. Keeping a slow hunch alive means to cultivate it: give it food to grow, provide fertile ground to take root and to link itself.

GP10 Grant time to grow!
Encourage the slow hunch: Many have their ideas in the workplace where they are exposed to stress and distractions, where there are superiors, whom they are accountable to and where they act under constant supervision. Improve the working environment so that ideas can arise and grow more easily.

Serendipity

Random links associated with a slow hunch lead to inspiration. Serendipity is when "knowledge and progress fly to us" when we find something we have not originally wanted. An artist of visual arts with one character appearing always and again in his artistic work in different forms and contexts once said to me literally: "The idea came to me after years of artistic activity."

GP11 Enable and exploit inspiration!
Open up the organization and create room for serendipity and inspiration: In our minds—as well as in the subconscious mind of our organizations—an almost infinite number of ideas and memories are slumbering, that can emerge at any moment on the surface and sometimes do this again and again. How can we get close to these ideas and exploit them? The challenge is to create an environment that favors those strokes of luck in the important areas: in our heads, in all kinds of organizations as well as in the information networks of our organizations and our society.

Error

Good ideas tend to occur in environments where there is a certain amount of disturbance factors and errors. One would believe innovation needed accuracy, clarity and focus. But the truly innovative environments are always a bit contaminated. Without faults (mutations) evolution would have stagnated and would have produced only perfect copies that would not have been able to adapt to changing environments. Benjamin Franklin once said: "Sometime we may learn more from a man's errors than from his virtues". The truth is uniform and narrow. Truth is eternal and to meet it you need less active effort but rather a more passive inclination of the soul. The error, however, is infinitely varied. In the United States it does not result in a stigma for an entrepreneur when he happens to lead his company into bankruptcy, but it is seen as a possible valuable experience.

GP12 Allow error and fuzziness!
Innovative environments also need "productive errors"—innovative environments suffer from too much control. Strict quality management methods such as Six Sigma and TQM try to banish all errors. They lead to bureaucratization; but they do not force conscientiousness necessarily. Web startups sometimes follow the motto "fail faster"—cope with errors as swiftly as possible, because errors are unavoidable, but we can learn from them.

Exaptation

Exaptation means thinking beyond boundaries of the respective subject or application. An example: The World Wide Web is a story of a coherent set of exaptations— Tim Berners-Lee developed protocols for a very special environment (for scientific data exchange in the form of hypertext documents). Today it is a platform for shopping, photos and videos. Sergeji Brin and Larry Page had the idea to weight the popularity of a webpage by the number of links that point to it (an exaptation of hypertext links as a measure for quality—originally serving as a navigation aid). The idea resulted in PageRank, the algorithm that made Google a giant. A slow "multitasker" with the knowledge of several disciplines in the head can be very innovative. "Chance favors the networked mind". Diversity and variety—be it cultural, be it by training—can be a success factor for the innovative strength of an organization.

GP13 Promote interdisciplinarity!
Build an innovative working environment in order to promote exaptation— for example as "interdisciplinary cafe". Encourage intellectual complexities, curiosity, hobbies, multitasking, a network of enterprises. Grant sufficient time for tasks, so they can also "echo" and thus remain in memory for a time. That may lead to ideas being born.

Platforms

The beauty and power of the simplicity of the Web 2.0 platform arises from the Web having the character of a platform: Web pages are hypertext documents that can be connected to other web content via a simple connection—the link. We take information that initially arose in a different environment and add them to the wealth of information now. The biggest advantage of platforms—one superimposed on the other—lies in the knowledge about the details of a platform that no longer has to be available for the acting individual, group or organization—or the platform above. This is the generative power of open platforms.

> **GP14 Reduce complexity by providing platforms!**
> Use open standards to provide platforms for the organization that can easily interact with the environment of the organization. Define intraorganizational platforms for tasks with comparable content—be it technological like ICT infrastructures, be it methodological or organizational platforms.

These innovation patterns repeat themselves on different levels. We may also create similar environments in our everyday life: at work, in the organization, by the way we use media or help trigger our memory—put together they form a consistent whole, which can be far more than the sum of its individual components.

5.2.2 Innovation by Forming Systems and Infrastructures

If we want to think organizations anew and we can see how innovation in technology is accelerating (see Sect. 3.2), it is natural to consider the innovations in the field of infrastructures of organizations first, and to then focus on the design considerations concerning organizational issues of new imagined organizations following the metaphor and the concepts of the subconscious mind of organizations and its interfaces to the conscious mind of organizations.

Let us recall once again the overall model, before we delve into the innovative design of the infrastructure in a deeper consideration. Let us look first to the embedding of the organizational unit in the external and internal spheres of influence (Fig. 4.5).

- Every employee has as part of his personal subconscious mind a "subconscious mind with professional reference" and it is integrated into the infrastructure of the organization and the environment of the organization.
- Each organizational unit and group has its organizational subconscious mind, which is characterized by its specific task and its infrastructure as well as its connection to the organization as a whole and its environment.

- The organization as a whole has an "organizational subconscious mind" and an infrastructure as well as a connection to its environment (Fig. 4.6).

Even if the top management of the organization is confronted mainly with this holistic view of the subconscious mind of the organization and will try to use it at its best, for top management one issue seems essential: Top management must also be aware of the specific subconscious minds of organizational units and—to a certain extent—of the employees in order to align the entire company as well as possible and to condition its handling of internal contradictions and trade-offs in decision-making at the relevant level in the right direction. So let us have a look at the technological innovations provided for innovating the organizations. Let us consider their potential impact:

5.2.3 ICT Infrastructure

If we consider the fundamental aspects of innovation following Johnson (2010), the most important pattern of innovation as basis for the design considerations of the information technology infrastructure is the pattern of "liquid networks" (see Sect. 5.2.1).

For innovation we need a "liquid network" of high density, as innovative systems have a tendency to chaos. A "fluid environment" indeed gives some stability, so that ideas do not dissolve or evaporate too quickly, but can grow to an innovation.

An essential basis of our organizations is knowledge. The basis of knowledge is information that can be linked in the relevant context and presented in a form so that the knowledge is readily applicable to a specific decision situation and that knowledge is easily linkable and associable for creative solutions. The basis of information, in turn, are the data-linked by the corresponding semantics—thus forming information.

In order to ensure an appropriate and sufficient "density of the liquid network" relevant for the organization and its services and innovation, an adequate knowledge-, information- and data-architecture is required. To be successful in the future as a "Real Time Enterprise" (see Sect. 3.2.14) will require flexibility and simplicity as the top priorities. This requires extensive reduction of the data architecture to "raw" data-linked to the specific organization in a kind of "Master Data Management" with clear semantics for these data. This allows the generation of knowledge in a dialogue with this information with the help of modern technologies in in-memory computing—something that previously was not possible. With the new technologies, the visualization of reality in a direct dialogue with large amounts of raw data is possible, rather than the former—and mostly still prevailing —monthly analysis based on complicated aggregated data. The previously necessary pre-aggregation of raw data for the often complex grouping hierarchies—in addition to the technology-related need to ensure a reasonable response time— induced these aggregate data structures and therefore complexity, for example in

ERP systems. The implications on the organizational innovation are described in the following chapters.

Equally the direct integration and processing of incoming data-streaming—is enabled. An example would be: For a customer just being serviced in an online purchasing process the system links already available customer data with real-time analysis to available data from similarly characterized customers and—with the right algorithms—the costumer is offered additional or related products. The real-time enterprise becomes reality—in real-time dialogue with its customers embedded in a liquid network of data, information, links and knowledge as well as easy to use and intuitive devices and applications.

Considering the developments in the industry, where the reality already can be promptly perceived by more and more sensors, we find that organizations cease from pursuing a costly increase in accuracy, which is often inefficient compared to the installation of numerous cheap wireless sensors. The effort to raise measurement accuracy more and more gets skipped due to the strategy that lack of measurement accuracy due to defective sensors gets statistically compensated by the large number of sensors—as long as there is no bias expected from the sensors being defective.

The data access is—with the new technologies—not any more "slow" due to the rotating disks, but goes directly and fast to huge working storage capacity with terabytes of data—the so-called in-memory computing. This huge amount of data can be (with new algorithms) processed quickly in a novel manner for novel and effective analysis and visualization. This also allows new—often more simple-application architectures with renouncing complex levels of aggregation of the data in order to rapidly obtain information.

> **GP15 Enable the "realtime enterprise" with simple architectures!**
> Simplify and make data-, information- and application-architectures of the organization more flexible by reducing levels of aggregation and concentrating on the processing of raw data with an excellent Master Data Management in the background. Take the chance of integration of current data streams in terms of "real-time enterprise" for the direct involvement of the customer as well as the reality of action in all relevant areas of the organization. Thus lift the "perception of the organization" to a new level.

> **Example hospital group—*looming sepsis detected earlier—prevent harm and costly compensation cases***
> **Paul, the CEO** of a hospital company had—after one of his hospitals had been sentenced to pay compensation to a patient who had lost his leg after a sepsis had been detected too late-, initiated an evaluation of ways to improve

the early detection of such disease processes. He had heard about a big data project in the US, where such a system with algorithms already had been implemented, where slight changes in patterns of various vital parameters in intensive care could be detected realtime—by comparatively interpreting these data with previous cases—whenever new vital parameter data were measured. That was about 2 h ahead, before a good intensive care physician —in view of the vital signs—could give indications of a looming sepsis. That could save lives or at least prevent severe harm for the patient. Considering the cost of a compensation case, such a system would give good returns quickly.

Example industrial company—*saving downtime of rolling mill by early indicating problematic vibrations*
Frank, the maintenance engineer in the rolling mill, had been just called by the project manager Gabriel. Gabriel had informed him that the bearings in the roll stand A7 should be replaced within the next 30 h. Gabriel had been alerted following the analysis as a result from the project with the plant manufacturer involved. Frank was skeptical at first, because according to his records about endurance data and actual operations data the bearings would not have to be replaced before another 150 h of operation. He went to the rolling line, trying to elicit whether there were any suspicious symptoms or strange noises around the roll stand A7 while rolling. He could not find anything. Since the roll stand would be taken off this rolling line tomorrow anyway due to the adapted rolling program, he ordered to have it dismantled first thing tomorrow. And indeed, the bearings already were in poor condition. They would certainly not have standed 150 h more. An unplanned downtime and sensible loss of productive time would have been the consequence. Frank was looking forward to the next meeting of the project team. He would urge those guys from the plant manufacturer to have him very carefully explained, what the patterns in the vibrations were that lead to the recommendation to disassemble the unit and what the exposure records and load characteristics for the last eight weeks of production on this roll stand A7 showed.

Facilitated with good visualization, the "dialogue with the data" allows new opportunities of generating knowledge—knowledge discovery. Especially referring to the hospital group scenario the integration of external knowledge resources, such as literature databases has high potential to improve professional work when you easily find relevant literature being in "dialogue" with the data of the actual patient and patients alike—the subconscious mind of the organization and the knowledge base available in this subconscious thus gets significantly augmented. But a prime goal will be to quickly and easily use the internal resources of knowledge for an issue or a question. This can be done in many different ways—from the statistical analysis of historical data and resulting predictions (*discharge management of Robert, the specialist for internal medicine*) through to find staff members with the necessary expertise via a good knowledge management system as part of the subconscious mind of the organization.

GP16 Provide easy to use context sensitive knowledge access!
Design a reasonable knowledge architecture with excellent knowledge management involving external and internal resources of knowledge and allow the user to get connected to the relevant data and information architectures with a good and context sensitive user experience. Thus the decision-making and business processes of the organization can run in a flexible and practicable way, based on the proper configuration for the organization's automated and subconscious as well as its conscious processes.

Examples for that already can be found in the previous stories and scenarios, for example, about discharge management in the hospital.

To enable these connections to external resources of knowledge and the use of internal knowledge resources, networks as well as computing and data storage systems are necessary, which support innovation in terms of a "liquid network" in the right balance of flexibility on the one hand and data protection and privacy on the other hand.

The necessary ICT resources can—at the current state of technology—increasingly be sourced from the "cloud"—similar to "getting the power from the grid" in electric power supply business. Cloud computing takes place either

- in the private cloud—an enclosed and secured internal area (for example, within the safe network of the respective organization), or
- in the public cloud (for example Dropbox, publicly accessible from anywhere but password protected) or
- in mixed configurations instead (in the hybrid cloud).

Meanwhile, there are powerful computers with the ability for massively parallel processing of the program code on many parallel processors—either on the same computer or distributed in the (now already sufficiently fast) internet ("Grid Computing").

The trend for "any time" use of the benefits of information and knowledge resources—at whatsoever place—requires a graded "data mobilization strategy", in which the safety classes are to be regulated in stages according to the requirements of the organization, for example, in the following grading:

- Which data/information are only accessible within the company, for example, certain customer information, employee information, intellectual property, etc.?
- Which of the data/information are also accessible mobile from a secure external network access combined with dedicated devices with appropriate authentication via token?
- Which data/information are accessible through authentication from any external PC or mobile device with user authentication on this device, for example, access to non-highly sensitive areas of the organization's intranet, e-mail accounts of certain employees, etc.?
- Which data are available to the public via the internet website of the organization like Open Data etc.?

In this context, a security architecture has to be defined and it has to be regulated as to which data are to be transmitted only encrypted. The questions are:

- What data are processed in which cloud (private, public or hybrid) or are they stored in different computer architectures,
- Which data, especially those who are accessible from mobile devices, are to be encrypted.

The answers quite significantly determine the security architecture of information and communication infrastructures including the network architecture. That should favorably be embedded in the holistic security policy of an organization.

Some examples: organizations cannot escape the network and the internet. If you imagine in the context of increasing cybercrime that for example the energy management of buildings including the shading systems of glass facades or the control of a sensitive plant or the heating control functions in homes are accessible from the vastness of the Internet, it is clear, that appropriate security architectures are required. Whether there will be—in case of failure—enough qualified and trained personnel for an offline or manual mode for these facilities and systems, will be seen. In light of the ever-increasing pressure on productivity and efficiency, I doubt that.

This shows two core threats and aspects of security:

- Unauthorized intrusion with criminal or terrorist background and
- Losing the ability to handle complex systems in a "manual" mode.

Both aspects have to be considered in a security architecture and in creating and living an overall security policy.

In addition to the liquid network as innovation pattern for creating a good ICT-infrastructure as foundation for an organization and as decisive part of its subconscious mind, it is wise to follow a second innovation pattern: the platform idea: platforms providing features, knowledge and security aspects within itself, that need not to be considered in detail from the platform or the application above. For example when you succeed to build your applications (the application architecture level) on only very few technology platforms where you manage data safety and security issues as well as performance management etc. in an excellent way, you do not have to consider this with every single application—you then only need to obey certain given rules. This prevents unnecessary complexity which often turns out to be risky. Thus you can easier succeed to provide "security by design".

GP17 Provide security and flexibility based on trusted platforms!
Create a security architecture that provides adequate security, but also enables the stationary and mobile use of information and knowledge resources as well as business processes from the standpoint of a positive user experience. Have your ICT infrastructures become a safe but also enriching element of "the subconscious mind of your organization"-with all its necessary diversity of equipment within the meaning of the "liquid networks" as well as standardized "platforms" as the basic patterns of innovative organizations.

Example hospital group—*ICT-security (by design)—working from at home—"there is no free lunch"*
Sandra, the chief physician of the internal department wanted to allow her employees access to the medical records on their tablets from home. She wanted this access in the simplest possible form: By activating the hospital information system—as they had access to their e-mails: turn on the device, unlock the device and go ... She had reactivated the unlocking feature only recently despite her desire for simplicity, after she had heard from friends about unpleasant incidents when their tablets were used by unauthorized third parties after they had left their tablets unattended for a short time. But Sandra wanted quick access when she was called by a member of her staff from the hospital, who needed her support from home. But she now had to dial in and authenticate with mobile tokens to obtain highly secure access to the patient data via remote access to the hospital information system. Obviously it was a comparable security strategy as her bank had with e-banking. That was annoying but—understandable—necessary. She now used this also frequently for literature searches related to complex cases with patients and she thus was able to pursue ideas that only came to her in the more relaxed atmosphere at home. However, she succeeded—at least one day a week—to completely "disengage". And emergencies, where her excellent team needed her counsel—yes it occurred, but not too often.

5.2.4 Social Media Integration

Probably the most relevant innovation pattern in this context is the exploration and opening up of the Adjacent Possible. There are systems and organizations that succeed better to explore new realms of possibility than others. Organizations with a good social media strategy are surely among those. It is necessary to look for suitable organizational improvements that enable these organization to accurately embed social media. Since such an embedding of social media can be expected in the long run on a broad front, the subconscious mind of the organization will change from a management perspective. The embedding of social media ("social computing") must therefore be managed proactively and consciously. It is not enough to just let it happen. It is wise to experiment with different ways to embed social media before the best solutions are strategically rolled out.

Oxford Economics (2011) thinks that the pure focus on conventional business transactions would mean a loss of knowledge for the organization. Cooperation between different generations of staff members must take place both (traditionally) transactional and in business context as a frame for successful knowledge transfer. This only works if cooperation and collaboration are embedded in the business process. Thus business processes increasingly come in contact with the "cloud". This requires "cloud-based" cooperation and coupling to the application systems of the organization itself—they need to merge (in hybrid clouds).

> **GP18 Embed "social computing" to "perceive" your costumers and "interact" with them!**
> Embed social media technology into as many processes of the organization as possible. Thus secure the growth of knowledge and the knowledge transfer by connecting with the world of the customer and the organization's stakeholders. Thus you can provide a subconscious impetus for cooperation of the different generations of employees within your organization as well as to the environment of your organization.

> **Example retail company—*How can costumer loyalty be augmented for a local retailer?***
> For **Mark** as a **marketing manager** in a retail company of the consumer electronics and household appliances industry it was clear that the sales outlets—with the advice offered there—as a sales channel increasingly competed with pure online sales channels like Amazon etc. His company was challenged to make better use of its more direct relationship with the customers and to create something like greater customer loyalty. The customer

now expected a branding and/or specific purchasing experience that was possibly tailored to them and consistent in itself: on the website, in e-mail marketing, in advertising, in the search engines, in the call center, in the Social-media forums and the aftersales services and assistance. Especially the integration of social media regularly was discussed with colleagues from senior management and occasionally led to fierce clashes. The corporate culture and the "subconscious mind" of his organization were not ready to open up without compromise and to allow a new cooperation with the customers in this new world of social media. It was probably a dangerous situation and a generational issue in the management, because the generation of 50- to 65-year-olds had great inner reservations and had in mind in particular the risks of such an opening.

The use of social media blended with the services and "physical availability" as e.g. a local retailer can be a niche to exist in. But old-style entrepreneurs will have to open up and change the subconscious mind of their organizations by adding new ways of costumer relationship management and sales channel excellence to the old traditional ways of "sales channel management" and provide a new costumer experience. Think of the saying: "Analogue will be the new Bio" using an analogy to often successful niches in local food supply chains.

5.2.5 Context-Sensitive Presentation of Information, Visualization

There are numerous sources of data, information and knowledge that are relevant for the organization. The first step is to identify the relevant data and to make them available for the organization. This concerns the internal sources as well as external sources, for example, from the connections of the organization on social media or from publicly available—paid or free—information services. A new increasingly important source are the so-called "Open Data": Governments and public entities make their data accessible for use by organizations of all kinds—except of course classified data concerning issues of high-security or data related to sensitive-personal information.

President Obama has—on the first working day of his tenure—instructed the heads of all federal agencies to allow public access to as much data as possible. At the beginning there were 47 data sources. In three years they were grown to 450,000 data sources available (see Mayer-Schönberger and Cukier 2013). Even in European countries we observe a similar trend—though this is a sometimes "tough" process.

The available data and information volumes are processable with Big Data technologies now available (Sect. 3.2.8) and also open up new options when (in terms of innovation patterns by Steven Johnson) "platforms" for the organization thus emerge. We can compose a "wealth of information" from information that initially arose in different environments. The great advantage of smart configured platforms for an organization is that it is no longer necessary to dispose of the knowledge within the platform. This is the generative power of open platforms.

Big Data makes probabilities from data. Peter Norvig, the "guru" of Google for Artificial Intelligence thinks that simple models and a lot of data are better than sophisticated models and little data. The insistence on exactness is a relic of the analog age. Only 5% of all digital data is structured. By allowing fuzziness, we open a window into a previously not accessible universe of knowledge—a quick view on general trends. Thus, the organization will begin to see the world from a much wider perspective than before and in turn a more complete sense of reality will develop (see Mayer-Schönberger and Cukier 2013). These statements by Viktor Mayer-Schönberger and Kenneth Cukier in their book "Big Data—The revolution that will change our lives" can be applied on organizations. The organization and its environment will thus present itself in a different perspective and—concerning the staff—the sense for the organization and its actions will be more comprehensive. The subconscious mind of the organization is about to change.

In order to make these opportunities available, it is necessary to offer and present these data, the information and knowledge context-related and on demand.

GP19 Explore the wealth of knowledge and data and enrich it!
Look for ways to provide external information and sources of knowledge in a context-sensitive way. Use the increasingly available opportunities of open data and other external data sources in combination with internal data sources in the context of the information needs of the organization.

Example Hospital group—*evaluating the outcome quality of hospital treatment with external data*
Paul, the CEO of the **hospital group** was repeatedly confronted with the complaint by doctors that, in view of ever shorter length of stay in hospital, the lack of feedback on the success or failure of treatment in hospital and the further treatment of patients at the general practitioner's office. Paul had now initiated as a first step to merge the death data from the Open Data from the civil registration office in the hospital information system. In the case of death within three weeks after discharge the case should now be discussed in the regularly held conferences on morbidity and mortality in the departments.

Mortality rates for comparable cases were now available from the national quality registers as well as relevant literature without investing much effort in searching. Thus for all medical OUs a substantially improved information base was available.

Access to the wealth of information and knowledge that is available to us today is mostly provided with search functions, which also make available information from unstructured data. These information- and knowledge resources are rapidly growing but not yet really "designed".

The art now is to make such access possible in factual, temporal and spatial context as well as user-friendly, while not restricting the view too much. The design of such context-sensitive applications is therefore a major challenge with regard to the facilitation of openness, innovation and efficiency. Thus the subconscious mind of the organization will be decisively influenced:

- What is proposed automatically,
- What alternative options are available,
- How will we be motivated to actively analyze and investigate,
- Where will the user simply "click through"
- How efficient and how focused on quality will we be working?

The right mix and design will support the following innovation patterns in your organization:

- To promote the exploration of the "adjacent possible",
- To stimulate the "slow hunch",
- To enable "serendipity"—the surprising recognition of connections and opportunities
- To support "exaptation"—the thinking beyond one's own area of responsibility.

GP20 Support your knowledge workers with context sensitive tools!
Prepare and present information and knowledge in factual, temporal and local context so that your organization can work in an open and innovative way on the one hand and it can work in an efficient and automated manner on the other hand. Sharpen the awareness of your employees and support them by providing well-designed applications enabling them to identify opportunities for innovating, broadening and deepening—also within the daily pressure for efficiency in the operational working environment. Grant them enough freedom to act.

Example hospital group—*new ways to present the medical history of a patient in a context sensitive way*

Sandra, the chief physician of the Department of Internal Medicine, found that the anamnesis of multi-morbid patients—due to the lack of context-sensitive processing of abundant data and an often problematic capability of the patients to communicate—was not satisfactory and the derived decisions regarding diagnosis and treatment were "in need of improvement". For some time now, new possibilities to access the patient's medical history had been made available, so that a quick overview—just for a chronically ill and multimorbid patient—was impossible especially for doctors who were not yet familiar with this patient. The latest innovation in the form of a word cloud with the diagnoses and therapies shown in different sizes depending on their relevance now offered new ways to work. The concepts for presenting the entire medical history to her doctors were dependant on the treatment context—for example, rehabilitation planning or acute care situations focusing on their respectively very different information needs. They could now delve into the medical history and they could quickly jump to the relevant chronological presentation of the relevant data. So they could quickly grasp the data relevant for the situation while the embedding in the overall context was not lost compared to the previous functionality. With the help of the "Word Cloud" the perspective could quickly be changed. For example they could quickly change the perspective from considering the development of the patient's cancer treatment to the consideration of his acute renal failure. Quality and speed had improved enormously. The computer specialists had explained why it was now sudenly possible in this speed: The data were all quickly accessible in the working memory. In-memory computing they called it. Immediately Sandra associated to the abilities of the brain and the way how the brain researchers explain the cognitive processes in the brain. Sandra had herself worked with the new system and found that the context-sensitive processing and "speaking with the data" supported and changed her type of cognitive processing of information and making her clinical reasoning and medical decisions. Her cognitive processes seemed to run quite different with the new system than in the traditional work environment. Everything performed more compact, faster and better—more intuitive and easy. Once these features had been piloted in her department, she had reported in the meetings of the chief physicians and her colleagues demanded a fast rollout. Paul, the CEO already complained that he had problems financing the necessary additional licences.

This shows how new access to the data can change the quality of decision making as the information presented in a context sensitive manner changes the "subconscious mind of the organization" giving access to more data and views an individual person cannot have available for her decisions on her own—not even the very experienced ones. So experience is enabled to grow faster within each individual person working in such an environment. That's also an effective way to better knowledge management.

When developing and implementing such new tools and views on available data it is wise to integrate the later users closely using collaborative and innovative methods and tools like "Design Thinking"—developed at Stanford and now widely spread. Bring together teams from different professions (like physicians, technicians, controllers, managers) and have them develop their future tools, processes and methods in Teamwork. This results in team-learning which itself changes the subconscious mind of the organizational unit or the project group and finally of the individual participant. It is the result and the "doing" that matters in successful innovation. It is crucial to find an "interface" to the individuals involved. That cannot be text and figures alone—pictures, animations (e.g. from simulations) in an excellent visualization are key.

GP21 Visualize your data and enter into dialogue with them!
Attach high value to the best possible visualization of the new "flat"—not hierarchical data-, information- and knowledge architectures. Make sure for the best possible user experience and the involvement and commitment of your employees and partners. Good visualization makes it possible to interact with the data in dialogue, to learn from it, to discover knowledge and to develop a sense and a feeling for what is going on and how to act best possible for your organization. This will decisively shape the subconscious mind of your organization.

Example industrial company—*develop a new sense and feeling for the rolling mill to operate and maintain it*
Frank, in charge of the **maintenance** of the rolling mill, and the **project manager Gabriel** discussed in the project meeting with the plant manufacturing company which plant and system visualization was necessary—down to the bearings of the roll stands, the marking of elements of the same type, the operating times, the vibrational patterns, the load patterns and the proposals for improved operational parameters—in order to enable meaningful analysis. Frank was determined to urge for the best possible visualization, since he felt that it was easier to understand images in increasingly similar

structure, presenting all relevant components and events along the timeline, and thus to be able to develop his "feeling" for his mill. The images and patterns in mind were important to him, since he could thus safeguard his experience, his feeling, his intuitive skills and his personal decision base.

Personal knowledge and the knowledge of the group and the organization is augmented with experiencing good, consistent visualization of complex issues and dynamics. That is effective learning and it will sharpen the intuitive skills of all persons involved—skills they will need in daily business and decision making and that will keep them in the driver's seat when it comes to closer collaboration with machines and Artificial Intelligence in changing organizational settings—in "hybrid intelligences" (see Chap. 6).

5.2.6 Decision Support and Decision Automata

Let's start with an example of how decision support might affect a customer, an example which may be close to manipulating the customer, as it happens in many Internet-based sales processes, in which the preference of the customer from earlier inquiries and transactions is already known. An exemplary scenario concerning the decision support for customers and the integration of customers would be the following: The costumer enters the shop of a retail organization, he is greeted by the staff and by a mobile application on his smartphone and guided through the store by the app on his smartphone with personalized offerings according to his preferences. The activation of the notifications within the app is done through the use of wireless technology with a short range, which is mounted directly in the shop.

Apart from retail stores this new form of interaction is used at trade shows and events, in the bank branch of the future and actually everywhere where a company wants to get in direct and personalized contact with its customers.

The brain researcher Spitzer (2012) describes in his book "Digital dementia" how the navigation system—and the gradual becoming accustomed to it—weakens the ability for spatial and local orientation. The same idea may apply on such a decision support system for the customer when they get used to it. Gradually the desire and the ability to watch and attentively validate one's own actions (System 2 —thinking slow) will fade away and the decision proposals will be perceived as intuitive suggestions of our system 1—thinking fast—and our subconscious mind. The character changes step by step from decision support in the sense of offering options to automated decision making. This also may apply on the above scenario in the retail store. This has a dangerous momentum as—if willingly or by malignant manipulative change in the algorithms—slight variations in the quality of decisions are introduced but are no longer perceived by the users.

With regard to the design principles of the subconscious mind of organizations I want to refer mainly to the guidelines postulated and backed by practical examples in Sect. 5.1.3.

In everyday decisions in a professional environment it is important to ensure that semi-automated decision-making processes, in which people are supported by decision proposals of decision support systems (DSS decision support systems), these people often see it as a relief from their responsibility as well as their pressure of time and they may "blindly" follow these decision proposals. Decision support systems should be designed so that the backgrounds of their proposed decisions are to be made reasonably understandable—at least roughly. We should be aware of the basic reasoning behind.

A special form of decisions is purely automated decisions. This purely auto-mated decision-making processes, which are indeed increasingly implemented, require good random-led review processes by (human) experts. Therefore it is necessary to document the reasoning from the algorithm correspondingly and traceable in order to ground the review on an appropriate basis and to allow a learning process and the improvement of the algorithms and their implementation.

In order to prevent the "blindness" or at least "refractive errors" of the decision makers in view of the decision proposals and the automation with the underlying—customizable and manipulable—algorithms, it is necessary to establish a learning and review culture, considering the sustainability and viability of an organization and not to sacrifice these to short-term efficiency and seeming quality improvement.

Possible measures against a habit of overly uncritical adoption of decision proposals may be:

- Plausibility checks,
- Case discussions,
- Training with simulators,
- Team Learning,
- Short stories, making the complex relationships tangible—"Micro Worlds" as described by Peter Senge (Peter Senge 1990) in his book about learning organizations.

Some considerations and explanations about the background of decision-support algorithms:

Basically Big Data makes probabilities from precise data. Better and more efficient algorithms are made possible by ever greater computing power. The increasing amount of data, in turn, improves the performance of the algorithms, who usually dispose of self-learning functions in the form of cleverly chosen iterations. Examples are automatic grammar checks and automatic language translation. Google has taken all translations on the Internet as a basis for the development of "Google translate"—Google translate arose—let's say—from the "flotsam of the Internet". "Simple models and many data are better than sophisti-cated models and little data," Google says (see Mayer-Schönberger and Cukier 2013).

A decision support system can tap into valuable correlations without knowledge of the causes—knowing what, without knowing why, is sufficient in many cases. A correlation quantifies the statistical relationship between two data points. A strong correlation indicates: when point A changes, probably the point B also changes. But there is no certainty, only a probability (see also quantum physics). So I can recognize at B how A behaves.

Correlations generate hypotheses—seeming interrelationships and connections. Thus special analytical precautions are necessary if you want to ground decisions on these hypotheses. A field of application is for example the rating of bank customers concerning their creditworthiness. One US insurance group linked, for example, lifestyle variables with health data of applicants—a kind of "profiling". This is probably at or beyond the limit of what the European Data Protection law considers admissible.

Classic science generates knowledge through a hypothesis based trial and error method. But personal and collective affections and preferences often affect the definition and the use of hypotheses as well as the choice of data sources. Now—in the world of Big Data—correlations generate hypotheses. So we do not necessarily need an accurate hypothesis today for a phenomenon in order to begin to understand our world. It is a data-driven approach rather than a hypothesis-based approach.

The classic decision-makers are thus always at odds with their intuitive desire to know causal relationships. They then often believe to identify causal relationships, where they do not exist. The result: you judge and decide intuitively, without to really thoroughly reason about the situation. But correlation analysis and other non-causal methods of reasoning and decision making are often superior to intuitive guesses about causes—as results of thinking fast (System 1). Moreover non-causal analysis based on correlations may be relatively quick and inexpensive. Correlations thus may show the way for causal research—and correlations are mathematically provable.

Data-driven decisions are thus either supplementing or replacing human judgment. They will let data speak for themselves. Statistical analysis is forcing humans to assess their intuition anew. Big Data and technology companies are—and should remain to be—only "toolmakers" and allow us a new view of reality. So it is all the more important to safeguard the ability of the humans to make decisions in an organization in a timely and convenient form.

The algorithms and decision tools of an organization should be documented and consciously maintained and not be left to individual "gurus". One of the challenges is to keep the management and control costs low. Self-control options are likely to establish themselves as self learning effects in the decision-making processes of the organization or the system. They should be consciously used wherever possible. A minimum level of competence for and about algorithms is necessary for organizations that want to shape their subconscious mind consciously and in a controlled manner. It makes sense for the organization to provide—in its handling of decision support and automated decision making—clarity and to set rules regarding what should be kept in the focus of awareness and what should be allowed to take place in the subconscious mind. The accuracy of the chosen strategy should be occasionally questioned.

It is an outstanding managerial responsibility in organizations to invoke the system 2—thinking slow—at the right time and sufficiently often without interfering with the efficiency. These managers need to understand the obvious organization as well as what lies beneath and beyond.

GP22 Sharpen the skills of reasoning computer generated decision proposals!
Create a learning environment and a review culture by questioning the decision support functions and the automated decision making processes regularly and critically in order to maintain and train the understanding of the employees and to ensure the timely intervention of the system 2 in the decision-making processes. Be aware of the necessity to improve and sharpen this continuously. For that the organization needs a minimum level of competence in algorithms and a safeguarded change process for decision algorithms. The visualization is particularly important in such a learning environment. The methods of "Visual Analytics" are to be employed in a targeted manner to support reasoning and traceability of decision making processes.

Example industrial company—*understand or get a "feeling" for the algorithms behind decision support*
Gabriel, the project manager, was—in his contacts with the plant manufacturer—specifically interested in how the systems could detect the looming technical defects in bearings, gears and motors from the vibration patterns and stress patterns. In the beginning this was a complete mystery. The responses of his partner always sounded very cryptic—it was statistics. Since Gabriel was mathematically reasonably talented and interested, he began to work with the algorithms—thus more and more deeply understanding the patterns behind and how to hypothesize the causalities behind. Gabriel as an engineer always wanted to understand the reasons behind. It was irritating him, not to be able to understand and comprehend decisions taken by the "machine".

Though not all employees and managers will be able to understand what is happening behind in those algorithms, you will have to look for ways to give them a feeling for example in decision simulators (see next sections) in order to shape the subconscious mind of your organization successfully.

Example hospital group—*evidence based medicine and decision support as danger for sustainable and lasting medical skills*
Sandra, the chief doctor of internal medicine, had long been monitoring discussions about evidence-based medicine (EBM). Some proponents wanted to adopt stricter policies supported by EBM. Sandra was ready to use the evidence-based guidelines, as it now was indeed impossible to keep pace with the state of the art only by reading medical journals and visit conferences— let alone transfer this knowledge and the essentials to her staff. Now the IT department of the hospital group wanted to implement guidelines in the hospital information system as decision support on behalf of the medical director. Sandra—as a member of the project team—wanted to fight for proposing alternative diagnostic and especially therapeutic advice at every stage but always to emphasize that the preferences and the living conditions and the situation of the patient had to be considered. The thought dawned on her that the young people would simply acknowledge the decision proposals for forensic reasons, but also from uncertainty and pressure of time and that they would thus gradually lose ability to critical reasoning and lose competence. She was also skeptical—due to her knowledge of how the medical and scientific community was functioning—, whether some guidelines and their underlying studies were not somewhat "interest-led" by the pharmaceutical industry. It was rumoured that many studies were never published due to results that did not sufficiently support the targets set out.

So decision support systems will possible turn to decision automata because there is the danger that their proposals are accepted without reflection. In this case the human will leave the driver's seat of the "hybrid intelligence". The subconscious mind of your organization might change automatically and unintentionally— without being deliberately shaped. In Sect. 5.2.9 we will deal with design guidelines for the subconscious mind of organizations from the viewpoint of organizational learning. But let us have a closer look to the potential impacts of simulation techniques before that.

5.2.7 Simulation

Innovation is always driven by the attempt to understand the future, to find out how to explore the "adjacent possible"—in the way Steven Johnson showed us in his basic patterns of innovation.

A traditional way to explore the future is—in addition to the predictive analysis as a result of the new Big Data technologies—the field of modeling and simulation (see Sect. 3.2.9). Simulation and related animations are ideal to create an understanding of and a feeling for causal relationships and the system behavior as a whole and is therefore an excellent training element. The most common methods of simulation are (according to Niessner 2014):

- DES (**Discrete Event Simulation**)
  By simulating individuals DES can easily be visualized and understood (for example, in the form of animations). DES is a standard for production, process and logistics simulations.
- ABS (**Agent Based Simulation**)
  The focus of the simulation is on the individual agent and its behavior and control (for example, autonomous transport vehicles, increasingly whole supply chains) or the behavior of the mass (in social science issues), general traffic simulations and traffic simulations at major events.
- **System Dynamics**
  System Dynamics is used for holistic analysis and simulation of complex and dynamic systems, for example systems with effect relationships difficult to quantify, complex and non-linear dependencies as in social, economic, biological and ecological systems, the spread of infectious diseases, simulations for strategic planning, adoption analysis of innovation effects, market developments, etc.
- Combination of different simulation techniques (**hybrid simulation approaches**)
  Often these simulation techniques are combined, for example, Event simulation (DES) combined to simulate a partially automated manufacturing system with agent-based simulation (ABS) for autonomously acting transport units such as cranes, autonomous vehicles (AGVs) etc.

Models and simulations thus are excellent tools for "Team Learning" and can communicate complex issues. The model is the medium of communication to share the aggregated knowledge in its full depth within the group or the organization. All assumptions can be communicated as a coherent model and in a harmonized way of looking at the key figures to top management, sponsors, boards of directors, investors, etc.

These classic simulation methods are based on relationships of causalities that are represented in the model and thus are causally comprehensible, for example as predictive models such as weather simulations or as epic, narrative models such as scenarios, sensitivity analyses, etc.

As instruments like a compass and navigators they sharpen our intuition. Models presented as stories and metaphors of the real world can explain sequences of events, focus attention to dangerous or critical situations, let us learn faster etc. The continuous improvement of business models and improving the decision-making are in the focus of these activities.

Correlation analysis with Big Data technologies and the causality-oriented approaches of classic modeling and simulation techniques should be used with a good combination of the two disciplines or by selection of the correct method depending on the given problem. For example, experts can—by analysis with Big Data algorithms like decision trees based on extensive historical data—recognize new causalities (Knowledge Discovery) and teams can augment and improve their knowledge and their sense of reality for taking decisions. On the other hand dynamic developments based on modeled interdependencies and their comparison with correlations from Big Data can help to understand the behavior of systems better in order to be able to manage disruptions and unexpected events in a better way.

GP23 Augment skills in simulation-supported team-learning!
Use the possibilities of simulation techniques—optionally in combination with Big Data technologies—in order to make predictions for developments. Experiment in team-learning processes based on simulation models with well animated and visualized user interfaces and thus try to augment the conscious skills. Then—advancing with simulation-based training in important areas—try to broaden the subconscious skills of your staff and your organization.

5.2.8 e-Learning, Gamification and Microlearning

Simulations and models are especially suitable to be presented in game-like structures and thus can increase the effectiveness of training by providing a positive user experience (e-learning with gamification). In that way, they penetrate the subconscious of the person concerned as well as the subconscious mind of the respective organization. This can be embedded in an e-learning environment as part of the knowledge management of an organization that supports ease of use and mobile access thus embedding learning processes in everyday work.

These systems have to include specifically built-in gaming mechanisms including reward systems e.g. in form of scoring, which are then compared with benchmarks, etc. thus increasing intrinsic and extrinsic motivation of the learning person. Such learning modules can also be used via mobile devices, if extended learning units are decomposed and then offered in correspondingly small "portions". Micro-learning for example is such a method, that tries to support a sustained learning via short sequences of learning backed up with sophisticated repeating algorithms—building on the results of the embedded testing of the trainee.

GP24 Enforce critical plausibility checking with e-learning!
Experiment with e-learning, gamification and microlearning to train decision-making situations and especially to train the ability for critical and intuitive plausibility checking of decision proposals from decision support systems.

Example Hospital Group—*decision simulators with built-in traps for training of medical staff*
Robert, a specialist in internal medicine working in Sandra's department, already worried for a long time how to train the young doctors, for whom he had the responsibility for—regarding job-education and training. In context with the increasing availability of decision support systems in terms of Evidence Based Medicine, he feared that they would succumb to the temptation to spare the deliberation process when taking decisions concerning diagnosis and treatment. Furthermore he feared that they would simply accept the proposal of the decision support system and—in this way—avoid liability if the decision was wrong—arguing that the proposal came from the organization being responsible for the system. He had, therefore talked to the software supplier and the IT department of his hospital group about building a "Decision Simulator", where doctors could simulate cases in working hours with less operative workload. They should be able to do this in the familiar environment of the hospital information system: e.g. train anamnestic analysis using the electronic health record of the patient, with simulated questioning of the patient and the simulated laboratory reporting its medical findings and with decision proposals from the system that could well be wrong. The involved young doctors thus were forced to constantly evaluate whether the proposed findings and decisions were plausible or wrong. His ideas were received with interest. Examples of simulators in form of gamification were already implemented in other industries. Users of these systems were eager to gain the best possible score and to get only little deductions for mistakes. There were at least signals for R&D funds to staff and finance such efforts and developments also for health care. In the long perspective, Robert—when he was old—finally wanted to be treated and medicated by competent medical professionals and not solely by decision machines. He considered the now available dummies in the Clinical Skill Center very valuable, where learning doctors and specialists could simulate difficult situations as part of their training. The instructor in the background sometimes built in traps while difficult and rare case

constellations were trained like an anaphylactic shock while inducing anesthesia. It reminded him of a flight simulator. Robert knew the addiction for gaming from his younger years when he was really good in landing airplanes in the flight simulation on his PC.

This way of learning and training on simulators can preserve and augment the skills of the individual employees in order to deal with decision support systems and decision automatisms in a reflective and critical way. Thus they contribute to the skills of the respective organization and are an important element and contribution to an organization developing towards a learning organization as elaborated in the following section.

5.2.9 Learning Systems and Learning Organizations

Very advanced forms of organizations—and it is an art of management to shape such systems and organizations—are capable of being learning organizations and learning systems and they are constructed and designed to improve themselves in terms of their purpose and goals. Finally they become "self-tuning" organizations (see also sect. 6.2.3). The ability to learn is also a prerequisite and a characteristic of intelligent systems and of advanced Artificial Intelligence. The empowering of individuals—provided their consciously accepting to take influence and to shape its analytical skills as well as its intuitive skills—is a crucial part of the design of adaptive systems and organizations.

If we as humans try to improve our skills and if we have enough room to practice and to exercise, we can augment our skills intentionally (consciously), but we train our subconscious unintentionally too. Thinking about learning organizations it is therefore important to address this issue of adaptive systems and learning systems at this point. In previous chapters we have already discussed numerous topics concerning new technologies in this context. In the synopsis, it is essential—in efforts to make organizations and systems capable of learning—to systemically understand these complex structures as a whole with all their obvious and less obvious interdependencies.

This requires, in addition to the best possible organization, which we will discuss in the following sections, to give the executives, the employees and the organizational units tools and methods at hand to deal with complexity and to understand this complexity—in order to subsequently also possibly remove unnecessary complexity.

In the section on the subject of simulation (Sect. 5.2.7) the formation of models had been addressed. This whole chapter about design of the subconscious mind of organizations is based on the model of an organization (Chap. 4) on a meta-level. We now consider a methodical approach to strengthen systems thinking, to deal with complexity and to learn at the individual level without technical aids.

GP25 Provide methodologies for system thinking!
Establish methods that enable staff members and managers, to illustrate complex organizational and mathematically elusive issues and to reflect their interrelationships and thus to response to external requirements, interference etc. Try to improve the systems and the organization itself with the goal to shape a learning organization.

One method—between mathematically oriented models and such meta models—to model systems and specially organizations in their mode of action is outlined below: The following mnemonics you should be aware of whenever making systemic considerations or—at an advanced stage—you should keep them intuitively and subconsciously present (following Senge 1990) Use it whenever one of the following conditions seem to apply:

- The problems of today are the "solutions" of yesterday
- Pressure generates counterpressure
- It is getting better before it gets worse
- The easy way out is usually back in
- The treatment can be worse than the disease
- Faster is slower
- Cause and effect are—in time and space—not always close to each other
- Small changes can bring big results—but the greatest leverage is often the least obvious
- You can have the cake and eat it—but not at the same time
- Sharing one elephant does not give two small elephants
- Your opponent or partner are not to blame—maybe it's your relationship with them.

Should one of these statements give the impression that in your present case for action the intervention of the system 2 could be wise, it is worth investing some effort and to give the situation a deeper examination.

The "basic vocabulary" of systems thinking, which can be appropriate in such situation, essentially is:

- The positive (amplifying) feedback with the result of accelerating growth or accelerating decline
- The negative (opposite) feedback with the result of stabilizing the course but possibly resulting in the forming of increasing resistance
- The delay, which substantially affects and determines the timing of the system— e.g. "Settling" in a steady state in an oscillating way or approaching a steady state asymptotically.

Here two simple cybernetic basic processes can be observed:

Fig. 5.1 System archetype
reinforcing circle diagram
(following Peter Senge)

Reinforcing Circle Diagram

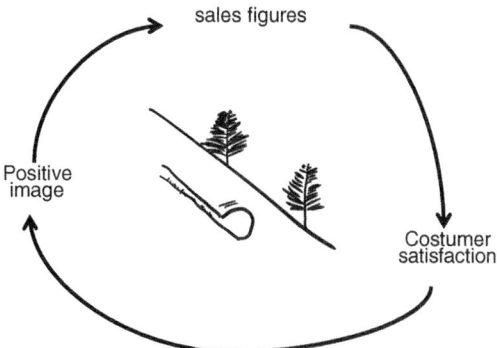

- The reinforcing feedback process, as shown in Fig. 5.1, where we can observe how small changes can grow either in desired directions or in vicious cycles where things start off badly and grow worse.
- The balancing feedback process is shown in Fig. 5.2, where we can discover the sources of stability and resistance.

Based on these elements and basic cybernetic processes complex systems can be modeled, simply thought through or even be simulated. When observing and analyzing complex systems, you can possibly recognize certain patterns, so-called system archetypes or generic structures.

Peter Senge succeeded to draft a collection of system archetypes from this "vocabulary" and to describe them clearly (Senge 1990). He has drafted these archetypes adding descriptions and characteristics, warnings and management rules for dealing with them as well as examples to illustrate them. At this point I want to provide you—in form of a short food for thought with loosely arranged "striking statements" to each archetype—the necessary basic understanding and a "first aid"—in case you encounter a specific difficult management situation. These

Fig. 5.2 System archetyp
balancing circle diagram
(following Peter Senge)

Balancing Circle Diagram – an example

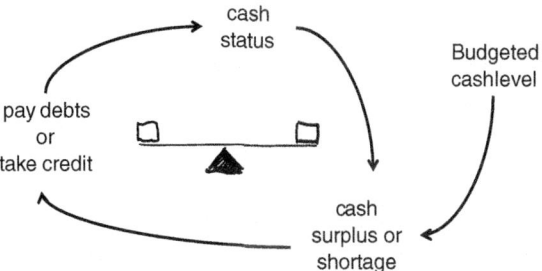

Fig. 5.3 System archetype
balancing process with delay
(following Peter Senge)

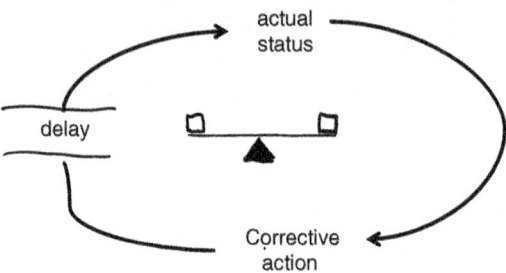

archetypes are intended to provide assistance for rapidly recognizing the situation you and your organization are in and to find the right measures and approaches. It is intended as a contribution to the shaping of the "subconscious mind" and its ability of being capable of learning.

Balancing process with delay

Behavior adjustment takes place only on delayed feedback. We are often not aware of the delays. The systems viewpoint is generally oriented towards the long-term view. That's why delays and feedback loops are so important. In the short term you can often ignore them. They only come back to haunt you in the long term:

"I thought this case would be in balance." "We have overshot the target." → Be patient and observe how the system really behaves (see Fig. 5.3).

Limits to Growth

A reinforcing process is set in motion to produce a desired result. It creates a spiral of success but also creates secondary effects manifested in a balancing process which eventually slows down the success. Thus the planned growth is delayed, stops or even tilts. The resources are limited:

"Why should we care about problems that we do not have?" "The more we make an effort, the tougher it becomes ". → Do not increase the pressure on the process, but be aware of resource limits and eventually remove factors that limit growth (see Fig. 5.4).

Shifting the burden (eventually to the intervenor)

The basic archetype "shifting the burden" means that a short-term measure or solution eliminates or mitigates a symptom, but does not address the problem itself. These are well intentioned easy fixes which seem extremely efficient. Thus the short-term measure or solution is being enhanced due to its success. In special cases sometimes the burden even is shifted to those who succeed with the short term measure—then you are shifting the burden to the intervenor:

"Look, it worked!" "Why do you think that we will run into problems later?" "Our clever management has intervened and will know what it does." "Let these

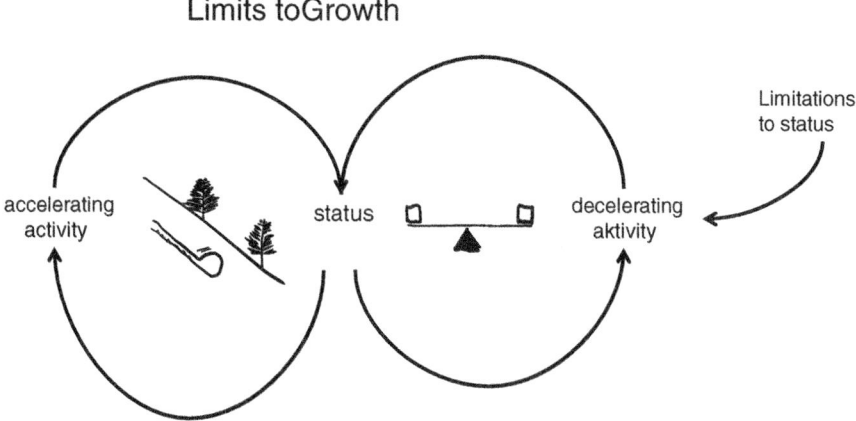

Fig. 5.4 System archetyp limits to growth (following Peter Senge)

consultants do their work …" → Work—at least in the background—on the fundamental solution. Teach the people to fish, instead of giving them fish (see Fig. 5.5).

Eroding goals

This is a shifting the burden type of structure, in which the short term solution involves letting a long-term and fundamental goal decline:

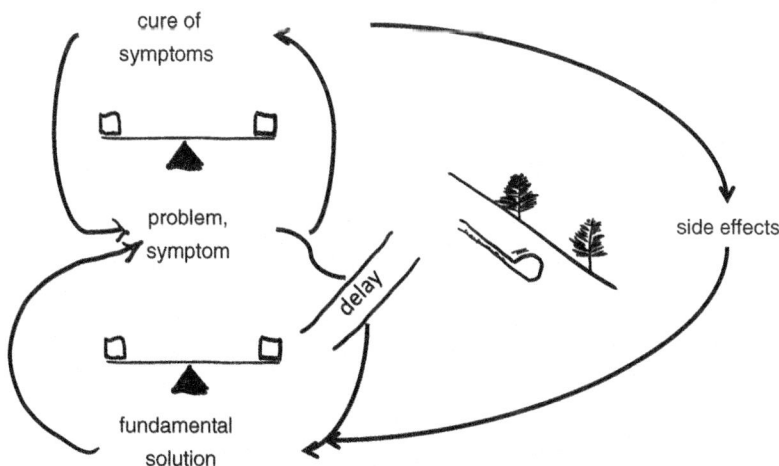

Fig. 5.5 System archetype shifting the burden (following Peter Senge)

Fig. 5.6 System archetype eroding goals (following Peter Senge)

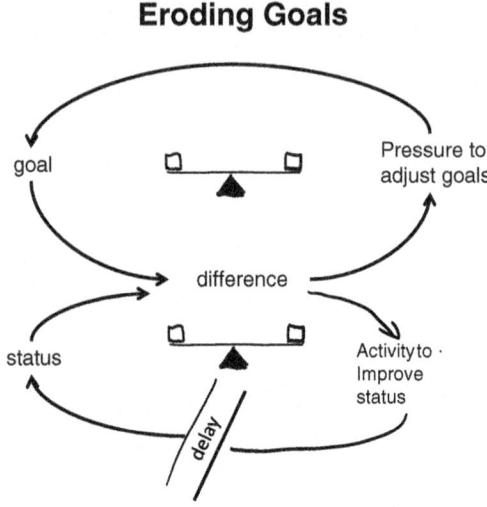

"We are tinkering about with the symptoms, but we lose sight of the goal!" "It does not matter if we take a more pragmatic approach, let us not see it too narrow— only until the crisis is over" → Keep the goal in mind—stay committed to your vision! (see Fig. 5.6).

Escalation

When two entities or persons see their benefit as depending on the relative advantage over the other this leads to an escalation of threatening and aggression— though often thought as being defensive:

"We are killing each other." Aggression is answered with aggression. "If our competitor would slow down a bit, we could stop fighting and take care of other things." → Is there a win-win situation? Shall we take out some speed? Can we make the other feel less threatened? (see Fig. 5.7).

Escalation

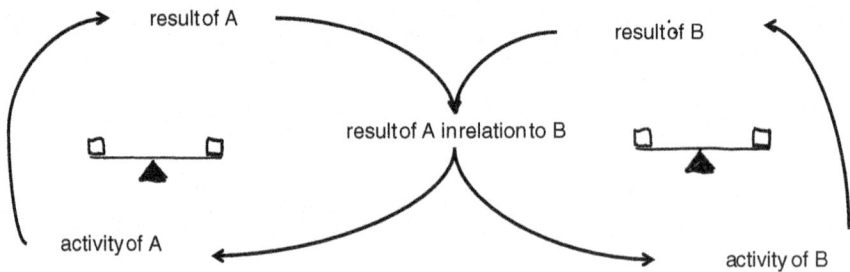

Fig. 5.7 System archetype escalation (following Peter Senge)

Fig. 5.8 System archetype
success to the successful
(following Peter Senge)

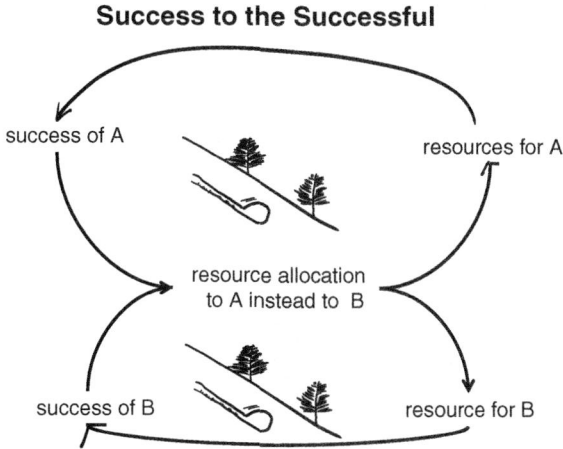

Success to the Successful

Success of the Successful

Two entities are competing or fighting for limited resources or support. The more successful one becomes the more support it gains, thereby starving the other who might have a tremendous long-time potential:

"Though the idea seems to be excellent, it seems to be starving"! "They have a great project marketing" → Look for the overarching goal. You also could break the link between the competing entities! (see Fig. 5.8).

Tragedy of the Commons

Individuals use publicly available resources solely on the basis of their individual need. First they are rewarded for using it, then diminishing returns causes them to intensify their efforts—until nothing is left:

"There used to be plenty for everyone! Now things are getting tough" → Implement self-regulating mechanisms, for example, by peer pressure, or ideally designed by participants (see Fig. 5.9).

Fixes that fail

A fix, effective in the short-term has unexpected medium- and longer-term effects:

"It always seemed to work before, why not now?" → Maintain focus on the long term. Short-term fixes should only be implemented to buy the time necessary for sustainable solutions (see Fig. 5.10).

Growth and underinvestment

Growth approaches a limit which only can be eliminated by investing in additional capacity. Often key goals and performance standards are lowered to justify underinvestment:

"We used to be the best and we will be the best again, but right now we have to conserve our resources and not overinvest." → If there is real potential for growth, important measures have to be set firmly and quickly enough to stay on the growth

Tragedy of the Commons

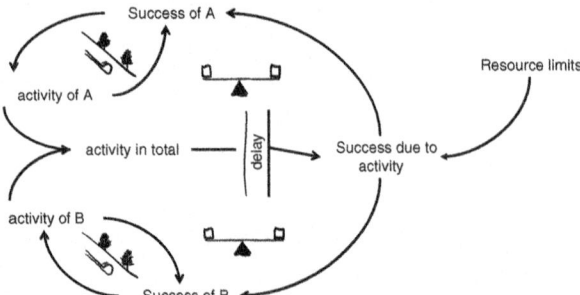

Fig. 5.9 System archetype tragedy of the Commons (following Peter Senge)

Fig. 5.10 System archetype
fixes that fail (following Peter
Senge)

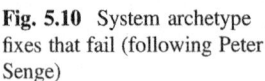

Fixes that Fail

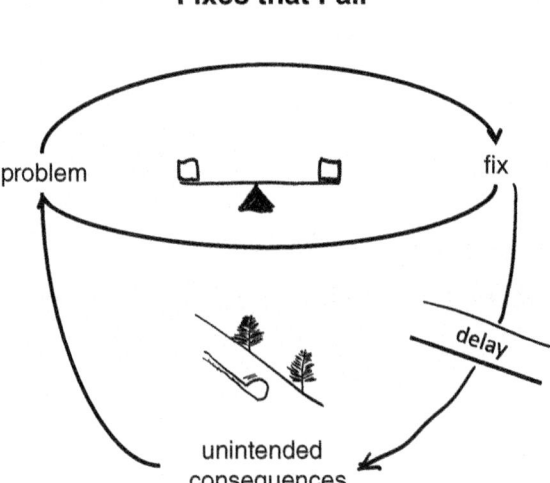

path. When it comes to a strategic enterprise you should invest in capacity in advance of the demand (see Fig. 5.11).

These archetypes are great for thinking through situations of decision thoroughly, to identify the system characteristics and to possibly outline the interrelationships of the situation as well as to incorporate possible developments in the model and then mentally simulate alternatives. Both the individual managers and employees who are regularly confronted with complex issues should internalize this kind of systemic thinking over time. Established teams themselves will show a more holistic and systemic behavior in their cooperation. In team training or following a routine meeting learning units may be incorporated in order to train this way of systemic thinking.

Growth and Underinvestment

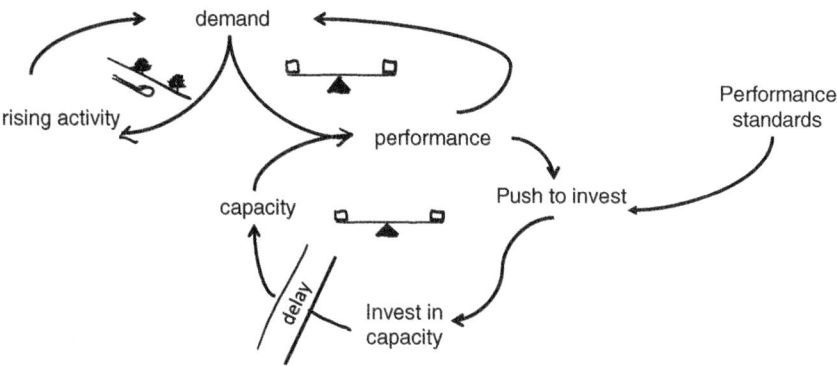

Fig. 5.11 System archetyp growth and underinvestment (following Peter Senge)

However, it is necessary to equip each employee with problem solving skills. It is wise to select employees who already have such skills and knowledge to lead such team trainings and "learning units" embedded in meetings and thus "improve the subconscious mind of your organization" or at least to "make it ready to learn".

The German grandmaster in chess Stefan Kindermann has presented his experience in the use of intuition in chess in order to apply it to decision making in complex situations in organizations. He calls this 7-stage decision model the "The Royal Road". Following Kindermann and von Weizsäcker (2010), this discipline of thinking for the individual can be described as follows:

- Get myself in top form and provide clarity with myself
- provide clarity in the environment—detect conditions of competition, affirm the actual state as such and accept it
- Generate creative ideas which make sense—if possible in the team
- Forward thinking—thinking "wide" and not too "deep"—think ahead 2–3 variants and simulate them—remain flexible after each new twist and possibly modify the plan or reject it.
- Precise definition of objectives—what is my key value?—What (not who!) is the "King" in my organization?
- Think backwards from the goal then—be not afraid, do not just think "defensive moves"—possibly "turn the chessboard"—what is it, the other is most afraid of? —change the perspective
- Reflection—what did really happen and why?—process possible setbacks constructively—organize the experiences and generate strength from that.

Gut feeling and mind must learn to work hand in hand. The intuition and the subconscious of individuals, the subconscious mind of teams, and thus the subconscious mind of the organization are sharpened and evolve in the context of the learning organization.

> **GP26 Enable learning from setbacks and deviation systematically!**
> Provide learning situations and a learning environment for the organization by skillful placement of methodological skills in thinking and self-discipline, by gamification- and e-learning approaches and provide a climate of dealing constructively with deviations and setbacks, so that sense and intuition of the employees and the organization can continuously improve.

Kindermann and von Weizsäcker (2010) also have created a checklist for the use of reasoning and intuition when planning and deciding. It provides valuable information on the working of the thinking fast (intuition) and the thinking slow (rational logic) by Kahneman (Systems 1 and 2). They can serve as a rule for "letting work" the subconscious mind of your organization as opposed to conscious reflection and conscious involvement of the appropriate management level:

Which circumstances mainly indicate to trust the intuition?

- In this area, I am/we are experts with extensive experience
- We have a very complex situation with so many factors, that the rational logic is overwhelmed—decide in the light of uncertainty
- Evaluations where the various social-emotional factors flow into
- Not much time to decide.

What circumstances indicate necessarily to challenge the rational logic or apply it as a check?

- Requirement of clear structure
- Emotionally charged risk scenarios and purchase or investment decisions—decisions with calculable risk
- Probabilities and general mathematical relationships
- Decisions in an environment where we are not experts
- Sufficient time resources for decision.

Capabilities for collaboration which are integrated in processes and the working environment such as discussion forums for collective groups or wikis, decentralized production of short learning or instruction videos, "sharing" of ideas and questions, as well as collaborative tasks and collaborative documentation provide effective learning opportunities outside the traditional formal learning environment. Thus learning employees, speakers and experts are enabled to exchange information, experience and knowledge at any time. This trend is also referred to as "social learning". Thus, for example, social platforms like wikis, team services etc. can be embedded in training settings, thus enabling cooperation and exchange between the learning people.

> **GP27 Foster collaborative reflection and cultivate storytelling!**
> Make sure that the organizational systems and decision systems run not only in routine operational mode, but also provide sufficient space for individual and organizational learning. This space can also be well used to develop the organization and its subconscious mind continuously. These are often only short sequences of collaborative reflection and discussions, but they represent very efficient learning opportunities-maybe resulting in illustrative "stories". "You do not always need a seminar hotel to learn and exercise."

Reflection and raising awareness of decision processes are of particular importance. Kindermann and von Weizsäcker (2010) believe that "we should first pay close attention to the signals of intuition and hold these. Then we develop a rational planning structure. Critical points should be identified (typically at the beginning and at the end of the alternative plans), where in turn the intuition is challenged and the rational logic may be considered as not competent. At the end of the process, the results are again to be submitted to the intuition—How is our feeling about this now? Ideally, a congruence or at least a close approximation between reason and intuition should be achieved."

An essential element of mankind's cultural evolution is the narrative which also holds an important place from the perspective of cognitive science—"story telling". Stories can be more easily internalized and affect attitudes and intuition, especially when they are consistent in the context of the organization and its history. Relationships and interactions remain as patterns in the brain by their recurring presence and reproduction and may also contribute—in the way of intuition—to analogics and other creative problem solving and decision making processes. Usually, only personal experiences are more formative. Compact simple stories of success—but also of crises and their coping—can thus not only be part of the consciousness of an organization, but by and by also become part of the subconscious mind of an organization. It is worth to focus on the "cultivation" of these stories as part of the overall corporate communications as well as the recurrent repetition and retelling by experienced staff members to new employees etc.

Even scenarios are a form of story telling—imaginary stories in the future. The stories in this book named "Example" tell such "stories" (hospital group, industrial company, retail company) trying to sharpen the sense of the reader towards the topic of this book to shape the subconscious mind of organizations in order to cope with the challenging changes and disruptions provided by the digital transformation, that we and our organizations are going through.

5.2.10 Automation, Robotics, Product Innovation

If we look at the relevant technological developments described earlier, it should be noted that in the field of Automation and Robotics (Sect. 3.2.13) dynamic—nearly exponential—developments gain ground due to the advances in Artificial Intelligence, sensor technologies, 3D-printing etc. These must be incorporated sensitively, wisely and individually in each of these technologies-using organizations. Since more and more happens automatically for significant parts of an organization—it happens "subconsciously" from the organizational point of view. So it is a challenge to specifically define what should be happening in the "subconscious mind" and what we should make us aware of from the perspective of the respective organizational level. Where should the specific decision-making take place?

The basis for the automation sector are "Cyber-physical Systems" (CPS). This has become a common term in the meantime. CPS describes the connection of sensors and actuators in production with the control and planning systems on the one hand and the electronically identified products on the other hand, bearing their "work plan" in themselves and knowing what processing steps they need next, and in which steps they are to be incorporated into that product. This partly happens using the Internet of Things (IoT) and this requires next to the communication facilities the right algorithms for self-control of such complex systems. Products and systems where product and service merge are appearing on the markets more and more. Examples include copying machines with a performance-pricing model (customers pay per page) or aircraft engines from Rolls-Royce, where since the 1980s, Rolls Royce can be paid per flight hours and maintenance is included in the fee (see Wipfler et al. 2014). In this respect it makes sense, if—as described in the example of the rolling mill—Martha as a manager of the steel company thinks about to sell the rolling mill to a partner with appropriate engineering and maintenance expertise.

> **GP28 Form semiautonomous systems to enable self-regulating structures!**
> Attempt to form semiautonomous systems—by using the option to store substantial processing information ("memory" and "knowledge") directly in the products—and thus to develop sustainable new production structures. Try to use these technologies to generate new products and services ("service enabled products", "Hybrid Product-Service Systems").

Thus production processes and products can be replicated easily and the production can be adapted quickly and flexibly to the respective requirements. Products can be manufactured more customized than previously. Entirely new products and services can be developed and brought to market. Robotics—its original focus being in the automation of manufacturing processes—has indeed developed towards semi-autonomous service providers and changed services themselves with the given tremendous technological progress. That allows new classes of services of hitherto unimagined quality:

- "Self-guided robots" (AGV = automatic guided vehicles) in warehouses,
- Minesweeper robots and drones in the military field,
- Household robots,
- Telepresence robots that allow a new quality of "tele-participation" in meetings in a hybrid physical-virtual mode
- Telemedicine robot that—accompanied by a general practitioner or a nurse— virtually bring the specialists to a patient's bed to jointly decide the further course of therapy—directly communicating with the medical and nursing staff on site and the patient and for example to determine whether a transfer to a specialized clinic is necessary,
- Surgical robots and
- Rehabilitation and Care Robots: Both in the postoperative acute care as well as in preventive care for preserving the patient's mobility "Robots with feeling and emulated empathy" seem to give valuable future prospects. The aim is to support the professional technicians. The robots feel the pain threshold of the patient and integrate them intuitively in the course of therapy and in the training of the patient, for example with special rehabilitation equipment for shoulders injuries after accidents etc.

In addition to new, flexible and customizable production processes wide doors open up for innovation of products and services or new combinations of products and services: so-called "service-enabled products." The organization can impress their customers and build up customer loyalty through better after sales services for example by giving tips for efficient use that go far beyond the reading of instruction manuals. There are more and more so called "embedded systems" in the devices, which are in the future increasingly accessible from the Internet (Internet of Things), which can also observe the way how the device is used and thus can make suggestions for improvements.

"Our problem is not to generate data or request more information" says the Futurist Eric Topol, a physician and director of the Institute for Translational Science at the Scripps Research Institute in La Jolla, CA., "but that only 5% of the data produced are analyzed and used". "The average car has about 400 sensors, the average smartphone 10, our body does not have any sensors." The increasing possibilities are not only changing the medicine, but also the laboratories. "We will still have a doctor-patient relationship (unlike driverless car), but most diagnostics will in the future be performed by or with the patient." Cal Tech is working on sensors to recognize the endothelial cells that can predict a heart attack in the body, and give notification to the smartphone. Similarly, in the future, cancer cells will be detected and early countermeasures will be taken. At Stanford University (see Topol 2014) not only the "gene expression" was stored in 1 TB, but also the epigenomic data (2 TB) and the Microbiom (3 TB) of people were stored in form of analyzable data.

Around these new possibilities and opportunities new innovative services and products will be developed, that for example will change the structures and the way how medicine is practiced today. The division of labor and the organization of healthcare providers will change. The organization, its infrastructure and the subconscious mind of the organization have to be remodeled or adapted accordingly. But the change of an organization and its control of processes and quality assurance must stay a very consciously performed task.

GP29 Monitor the costumer focus of your organization closely!
Design and build your products and services so that maximum customer benefit and customer loyalty result and endorse product development, production and service processes in the organization in a way that the customer benefit and the self-interest of the organization are well balanced. Customize the controlling and monitoring processes at the management level as part of the "conscious mind of the organization" being responsible of the monitoring and controlling of the organization and—on the other hand—provide that the routinely performing customer-, service- and organizational processes are largely taking place in the subconscious mind of the organization.

Example industrial company—*adopted engineering processes including costumer's data raise costumer loyalty*
Ralph, the **operations manager of the rolling mill**, now already had some experience with the new simulation and optimization algorithms and the systems of production logistics from the steel mill to the rolling mill to heat treatment and to the finishing of products. The individual semi-finished products were all clearly identified and already "knew" what their next manufacturing step was and what process parameters such as rolling temperatures and deformation steps were necessary here. The troubles of rescheduling in case of a production change or disturbance of production were low—much lower but with incomparably better results than before. The decision proposals could be vividly made plausible in simulations and then the best decision could be taken. The process parameters of each semi-finished product were documented. The experience of the customers with their products "downstream" for example processing them as tool steel in mechanical manufacturing, etc. the production process parameters were again made available for the steel company and now could be analyzed together with the production parameters at steel and rolling processes of their company with Big Data technology. These data were analyzed by the in-house metallurgists and—where appropriate—the results were discussed

with the customer. This led to improvements in the process parameters. Thus the steel products could be produced more precisely and economically. Ralph knew that the customers knew that they could get such quality tool steel for their specific applications only with his company. The loyalty of customers was higher than ever before with these new additional engineering services. The process engineers and many of Ralph's employees now had more contact with customers than ever before. Martha, the managing director of the steel company, noticed that these extensions of the product by services not only increased the loyalty of customers, but the company itself was increasingly influenced. The culture and the subconscious mind of their manufacturing company had changed.

Example Retail Company—*"back on track" with multi-channel excellence combined with local services*
Mark, the marketing manager of the retail company last year had submitted the board several proposals to move towards "Smart Commerce" and better "costumer experience". The board had reacted with surprising speed. They already had been in economic problems. Now first results and successes appeared. The connection to the market and to customers had been designed anew. The "customer's experience" with incompetent sales staff, where you expected good counseling in vain, with hotlines where the costumer mostly was further connected after a long wait and with annoying calls from online sellers etc., had been successfully avoided after the first changeover period. They were now able to give the customer a brand and purchase specific "experience", which was consistent in itself: on the website, in e-mail marketing, in advertising, in the search engines, in the call center, in the social-media forums and with the services and assistance after the sale. Only the ranking in the major search engines was not convincing. They were now able to utilize the strength of sales outlets as a consulting facility, as a repair- and waste collection facility and they had established online channels. Many potential customers in the region now had Mark's retail company as favorite on their PC or Smartphone. So the ranking in Google was no longer of importance. The local cooperation with craft shops and the logistic department's performance in the new distribution center with its warehouse robots etc. was impressive. The management was optimistic that they could reduce the debt ratio again soon, which was high due to the investments in ICT and logistics. The customer orientation of the employees, the quality of advice and counseling for the costumers in the shops and on the phone were great.

The existence of the company seemed secured. They were "back on track" again. Mark was convinced to have made an important contribution.

Example hospital group—*excellent reputation and cooperation with partners as well as engaged patients*
Sandra, the chief doctor of internal medicine, was talking with Robert, her longtime senior doctor at a dinner on the occasion of his 30-year service anniversary in this department in a relaxed atmosphere. They reflected the rapid development of the past three years, in which so many technological progress had happened, with opportunities they had not expected at all five years earlier. Big Data, the new possibilities of sensors and diagnostics and personalized medicine had changed dramatically. The services for their patients had become different. The former practice-based specialists for internal medicine in the region were now mostly active in the hospital—many of them freelancing—and rendered their outpatient services to the patient at their location. The hospitalization rates had come down.

Many services were now provided via telemedicine, e.g. teleconsultations for general practitioners or geriatric counselling for nursing homes. Most patients' engagement had changed—the measuring of simple parameters like blood pressure, weight or blood glucose as the basis for diagnosis and subsequent monitoring took place primarily at home and was visualized for the doctors in charge—especially for the many chronically ill patients with cardiovascular, cancer or metabolic diseases.

Difficult cases could be now well prepared with the new context-driven links to literature and knowledge databases. The monitoring data such as blood pressure, blood glucose, weight, blood coagulation parameters etc. were clearly visible in their development on a timeline. Even biomarkers of impending heart attacks in patients at risk, which could be detected with implantable sensors, appeared on the horizon of the medical progress.

Decision support systems made unobtrusive suggestions for diagnosis and therapeutic procedures, they gave hints for the clinical reasoning behind, but actually they were mostly plausibilized by their staff members. The new training methods on the simulator had worked and were now gladly accepted. The conversation with the patients could now be performed on a much more individual level. The patient now had a "main medical contact person in the hospital" as it previously only was possible in the surgery department. The management of appointments for the medical examinations was much better now. In emergencies, the respective doctor on duty was informed quickly and had a comprehensive context-sensitive view on the patient's medical history.

The administrative workload and the search effort for the physicians had significantly decreased. The follow-up of patients was often carried out with an "email visit" with structured questionnaires, so that all doctors knew much more about the results and the success of their treatment than ever before. Sandra and Paul felt that the quality had improved and their doctors felt more secure in their decision making and better in their working environment. The working atmosphere and the culture of cooperation and communication had improved. They could now select from many candidates for apprenticeships. Their shaping of the subconscious mind of their department seemed to be successful. The reputation of their department was excellent now.

The increasing costumer and patient engagement has a high potential not only to change interconnectedness of the business, but to provide high potentials concerning costumer loyalty and to bring more and more staff members in contact with costumers, thus changing the self-motivation for staff members and providing costumers not only digital costumer experience but also positive analog experiences. This requires very careful and well orchestrated shaping of an organization which in the end may result in quite significant and positive changes in the subconscious mind of the respective organization. One cornerstone in such augmented costumer relationships is the trust of the costumer in the organization. This leads us to the question of security issues concerning an organization and its infrastructure.

5.2.11 Security

The extensive use of new technologies and their integration into products, services and processes that are most essential for staying competitive in the organizations' respective market and that is part of the subconscious mind of the organization requires a conscious approach to the issue of safety and security in a manner that innovation and creativity are not compromised. In addition to the discussions on security architecture concerning the information technology infrastructure, supplementary considerations are to be made at this point: In the weekly magazine ZEIT (see Himpsl 2014.) the subject of security—"digesting" the NSA scandal—was well summarized with a look at the industry: Networked, highly automated factories offer many targets of attack. The attacks do not come only from homeland but also from foreign competitors—and intelligence services. A survey by the auditing company Ernst & Young found that the fear of attacks from abroad has increased significantly. Especially mistrust against the Americans has grown: 26 percent of German companies—surveyed in Summer 2013, stated that the United States posed a particularly high danger concerning industrial espionage—two years earlier these had been only six percent. The Trump presidency does not seem to calm down this eroding level of trust (as of spring 2017).

The problem of industrial espionage has become more "explosive" mainly because a large part of corporate communication today happens digitally. And complex IT systems have vulnerabilities: Wireless corporate networks can be hacked, firewalls can be outwitted, mobile phones can be listened to. An infected e-mail attachment can, once opened, spread on the computer and draw information to the attacker. The risk of espionage—but also the deliberate damage by terrorism as well as blackmailing with corresponding threats—is particularly large for the factories of the future ("Industry 4.0"), where the production and assembly operations are negotiated independently between the machines and the workpieces with each other—an Internet of Things. In a conventional plant it is relatively hard to get from the outside to the assembly lines and manufacturing plants. In a networked, highly automated factory, an attacker could inject external viruses and shut down the production. The CIA had supposedly destroyed with STUXNET the uranium centrifuges in Iran's nuclear facilities, at least partially.

IT security is becoming increasingly important for all businesses. Often sensitive corporate data will also be lost without an external attacker nestling with great technical effort in the digital infrastructure of an organization: Against a bribed informer, the smuggling out of copied drawings etc. an organization is hardly immune.

The espionage risks through its own personnel are underestimated by many organizations. The risk primarily is based upon notoriously disgruntled employees, from people who are frustrated and have the feeling of not getting their career opportunities in business. Any organization who wants to protect itself from industrial espionage has to seriously deal with this human factor. High employee loyalty is one of the most important safety precautions. Security is also an issue of working atmosphere. Options for the determination of the status of threats are anonymous staff surveys or professionally provided "Whistleblower hotlines" where anonymous hints may be provided from everyone in the organization: You have to find out where there are interpersonal problems—and then you also have to consider consequences. Often the mentality of an entire organizational unit can be improved significantly by replacing the individual executive manager. It's about "psychological safety at work". It is crucial especially at critical points to assess the self-esteem of the employee—as it also generally applies in corporate health management. Let us assume that the self-esteem consists of three elements: family/friends, profession, the person themself (self-confidence, health, etc.). If one element makes more than the sum of the other two, the risk of causing troubles is high in case this one dominant element breaks away. Then with this person (following Müller 2014) the "psychological safety at work" is possibly at risk and the risk of deliberate wrongdoing is high.

Another risk to be addressed is data privacy. Breaches of data privacy e.g. of costumer data or patient data have the potential to damage your costumer's trust significantly. One major issue therefore is to raise the awareness for data privacy and data security and really implement this in the subconscious mind of your organization e.g. by videos combined with tests (supported by gaming elements—see above) for security issues and compliance issues like data privacy, data security, corruption etc.

> **GP30 Raise security awareness without losing the culture of trust!**
> Build a wise security architecture which takes into account the technical, organizational and human possibilities to err and define clear responsibilities for these elements of security within the organization. Raise the awareness of your staff, communicate adequately and openly—without compromising security—and avoid to establish a culture of mistrust.

Things you can see, you can understand and you appreciate are more easily manageable than things happening in the background and clandestine like the above mentioned security threats and safety issues. Something hard to understand for the majority of people and employees—and therefore dangerous concerning trust and manageability—is Artificial Intelligence and its impact on the subconscious mind of organizations.

5.2.12 Artificial Intelligence

The topic of Artificial Intelligence was highlighted earlier concerning its basics, but also concerning implementation issues as well as from philosophical viewpoints (see Sect. 3.2.15). What does this mean for the "new thinking" and the innovation of organizations? What is already relevant?

IBM has unveiled its chip True North (Modha 2016), which operates with low power consumption (0.1 watts), 4096 synaptic nuclei (which equals one million programmable neurons with 256 million programmable synapses). This required up to 5.4 billion transistors. The design is an "event-driven" design. True North stores data, processes it, and each neurosynaptic core can communicate with each other via a crossbar communication architecture. In comparison the human brain consists of about 100 billion neurons and 100 trillion synapses. The special abilities of this architecture are pattern recognition and classification of objects. "He could become a control center for the Internet of Things and completely change the mobile experience, as we know it now", the chief developer of the area "Brain inspired Computing" at IBM, Dharmendra Modha says.

These chips—also competitors of IBM are working on such designs—are opening the door to "Neuromorphic Computing" and enhanced "cognitive computing". In order to achieve higher efficiency and better mimic the many connections in the brain, "stacked" chip designs will be necessary.

These technology approaches now available with Neuromorphic Computing will feed into the two major projects for the research and for "replica" of the brain: The Human Brain Project (HBP) in Europe and the "Brain Initiative" in the US.

Richard Frackowiack from HBP believes that—in a medical context—we are on the way from the symptom-based diagnostics to mechanistic diagnostics and that diagnoses catalogues will have to be partly rewritten. The objective of the HBP is to construct and to build a blueprint of the brain: in all dimensions from the "base pair" to the cognitive level (see Frackowiack 2014).

The disabled researcher and prosthodontists Hugh Herr of MIT has developed a prosthesis for the ankle, which controls the motor and the spring in the artificial ankle via intelligent electronics and software with—in prosthetics already widespread—impulses from the leg stump. This prosthesis, and numerous other prostheses available on the market were assembled to a robot and an android with amazing abilities in a project by a team led by Richard Walker (see Mayer 2014). It is striking that in the prosthesis, distributed intelligence is also incorporated and not everything is controlled by the "brains" of androids—like with humans. Even in systems with advanced capabilities of Artificial Intelligence, the architecture of distributed intelligence will probably prevail for some time—this feature will support the analogy of "intelligent organizations" or "organizations composed of humans, hybrid intelligencies and purely artificial intelligencies" where humans and machines closely collaborate (see Chap. 6).

The Austrian philosopher Peter Kampits (2014) has noted in the Vienna Lectures that it is an anthropological constant of human nature to always want to exceed the natural conditions. This "accompanied" evolution through millions of years. Now these aids are becoming independent—we can no longer dominate them. "The human wants to go beyond what it is".

The cultural scientist Karin Harrasser in her book "Body 2.0—the technical expandability of humans" (Harasser 2013) speaks about media as an infrastructure of thinking, feeling and acting, as well as about the pressure that is developing—increasingly followed by many people—towards the possibility of self improvement—more and more establishing the imperative of self-optimization. The basis of her research and her considerations are her observations on the development and application of prosthesis in humans.

These considerations are relevant when we think about how we can make arrangements in the organizations how to use Artificial Intelligence, automated decisions, self-learning and self-optimizing systems, algorithms and processes effectively—from the viewpoint that, what is technically possible, eventually finds its application. The technologists are already far in the intracorporal collaboration of human and machine: nanotechnology, microelectronics, the ability to implant chips etc. appear feasible—including the ability for example to control the aggressiveness of soldiers. In border areas legislature already has been established—for example in Austria: Eugenic practices and human cloning are prohibited.

Systems with capabilities of Artificial Intelligence will be able to independently interpret content. Search engines will lose their relevance and will be replaced by something like knowledge machines that give situation- and context-specific recommendations for action, and if necessary will be offering alternatives. Siri from Apple, Alexa from Amazon etc. are already widely used chatbots as special form of AI.

How do we position ourselves in our organizations and organize so that we,—organizationally and ethically aware—can deal wisely with these features of Artificial Intelligence etc. What shall we keep in the background—in the subconscious mind of our organization? Much of this "Artificial Intelligence" will "creep in" there. We are called upon to recognize that and in terms of the conscious design of the subconscious mind of the organization properly use these technologies. More about the coming collaboration of humans and machines you will read in Chap. 6.

GP31 Monitor the learning of AI-systems, scanning for unintended side-effects!
Evaluate the possibilities and scenarios for the use of Artificial Intelligence in the form of (partially) automated decision-making and of self-learning as well as self-optimizing systems, algorithms and processes thoroughly. Implement processes to evaluate the learning progress in terms of improvement towards the goals set. Carefully observe the possible negative and positive side effects. If the changes are not plausible, a conscious reflection before implementation and continuation of the innovation is indicated.

Example hospital group—*keep self learning algorithms under close watch and quality control*
Sandra, the chief doctor of internal medicine, had realized that the rule-based system for measurement and administration of blood glucose and insulin parameters with ICU patients and other patients being in use for years should now get enhanced with a self-learning function. A self-learning function, which should optimize the insulin dose delivered based on statistical Big Data algorithms from hundreds of thousands of cases with their effect patterns from insulin doses delivered in certain medical situations. Sandra first stopped this process and consulted with her staff, how they should deal with this. Although this software was certified according to the Medical Devices Act, she did not want to rely unconditionally on this automatism in her responsibility. She called for a monthly report on the progress of self-learning algorithms for a year in relation to 50 standardized cases in order to learn how the results from the self-learning algorithms improved and changed over time. Furthermore Sandra confronted the system with 20 medically plausible case variants, arbitrarily determined by her employees. The results from the algorithms were discussed and checked for plausibility

with several staff members and a specialist in endocrinology. This would—acting upon her ultimate responsibility—strengthen the quality assurance and provide a learning effect for the employees. They should still be able to determine the insulin doses themselves without the system properly sizing the doses.

Now the innovation itself and the innovation of infrastructure have been covered extensively and ample evidence and hints have been given for the design and shaping of the infrastructure of organizations—constituting a decisive part of the subconscious mind of organizations. Now we want to expand the design considerations and recommendations on the organization as a whole.

5.3 Thinking Organizations New

If we look at the model of an organization (see Chap. 4), it describes a hierarchical structure. How deep or how flat shall we design the structure of an organization in the light of new technologies and from the "behavioral psychology" and "cognitive science" point of view on the subconscious mind and the conscious mind of organizations as emphasized in this book? This will now be elaborated in the following sections.

These considerations are based on the former considerations on innovative infrastructure and on linking the organization to the sphere of internal and external knowledge sources—all induced by new technological possibilities. This sphere of external sources of information and knowledge often gets little attention—especially at very self-centered organizations. Sometimes the resulting opportunities and effects are not recognized as such. The even closer integration with the environment and expansion of perception by the Web 3.0, by tools of Augmented Reality and by the Internet of Things—the Real Time Enterprise—offers new, still largely unexplored possibilities: Opportunities for innovation for the agile and adaptive organizations, but perceived threats for the conservative and immobile organizations.

The connection of the organizational units and their staff members is provided partly directly to these "external" spheres. In any case they should be well connected to the sphere of internal information and knowledge sources, which in turn are part of the technological infrastructure of the organization (see Sect. 5.2.2).

The performance of search engines and "knowledge engines" and their manageability and accuracy concerning their hits (often perceived as "usability"), which should ideally be constantly improved for the respective target group by learning algorithms, are of particular importance as data and information are stored increasingly unstructured—sometimes they are at least "tagged".

A basic feature of successful organizations of the future will be the collaboration in terms of "working together" and "sharing".

5.3.1 Collaboration and "Shareconomy"

Collaboration within the organization and with partner organizations and the sharing of resources and knowledge are essential elements of an economic paradigm of sharing. This economic paradigm of sharing ("sharing economy" or "shareconomy") is more than and different from the paradigm of an economy based on division of labour and competition, as we know it since the beginning of the industrial organization. It was the new technologies that have pushed open new doors and created new business models. That this sometimes can be perceived as a threat to established industries in certain segments can be observed for example with renting apartments to tourists (for example Airbnb), with car-sharing especially in large cities (for example Uber) and the organization of journeys and transportation services (for example www.checkrobin.com, an online platform that allows individuals simply, quickly and flexibly to transport all kinds of things or themselves).

Collaboration is more than cooperation and requires more willingness to share. A key element here is the sharing of knowledge. An efficient market and efficiency are the benchmarks for each business and for each economy—except when it is about ideas. We have created inefficient markets, when it comes to innovation: copyright, patents, trade secrets. This does not mean that patents should be revoked and all forms of innovation should be unprotected now. What we see, according to Johnson (2010), a researcher on innovation, is: "The creed that—without the artificial shortage generated by intellectual property rights—innovation would come to a halt, is simply wrong. If we ask the question of which system is more natural, the free flow of ideas easily outrivals patent law and resulting artificial scarcity. Unlike food, raw ideas are self-reproducing by nature". Lone wolves that elaborate patented innovations in their laboratories (as Alfred Nobel) are rare. Collective inventions are no socialist or socio-romantic myth. The freedom to be able to rely on others for ideas, often brought more benefits than the secrecy. Without the slow hunches and the ideas of the competitors the really big steps towards the adjacent possible often might have failed. "Many good ideas came even where any financial incentive was missing," writes Steven Johnson.

The "Open Data"—movement and its results are based on the free access to Big Data as a basis to generate new knowledge.

The collaboration is also facilitated by new standards in process integration across organizations. This is a significant advance over the standards for pure data and information exchange as developed in the 80s and 90s of the last century. One example in the health sector is IHE (Integrating the healthcare enterprise), where the process profiles for standard processes and special processes, for example, in the field of cardiology enable better collaboration. IHE as standard is already obligatory in public tendering in some countries. This will, for example, urge medical device manufacturers to become more interoperable, thus enabling cross-organizational processes for patients at an affordable price.

As a basis for the "rethinking of the organizations" you will find in the following sections some remarks how to shape the culture and the awareness of organizations. Both are very much influenced by the organization's subconscious mind. The ability to collaborate and to share are indeed a question of attitude and culture. Besides, leadership and management issues also need to be addressed.

After that we will consider first, what we have to focus on, to innovate the processes and the process organization in order to make the subconscious mind of organizations as fit for the future as possible. Then we will highlight the opportunities in designing and engineering the organizational structure.

5.3.2   Culture and the Conscious Mind of an Organization

Following the business dictionary (Business Dictionary 2017) "Corporate Culture" is defined as followed: "The values and behaviors that contribute to the unique social and psychological environment of an organization.

Organizational culture includes an organization's expectations, experiences, philosophy, and values that hold it together, and is expressed in its self-image, inner workings, interactions with the outside world, and future expectations. It is based on shared attitudes, beliefs, customs, and written and unwritten rules that have been developed over time and are considered valid."

Culture therefore also is a key determinant of the subconscious mind of organization as it conveys what is appreciated, what is a no-go, what is rewarded, what gets punished and thus leads to behavioral patterns depending on the congruence of the individual's values and attitudes with those of the organization. Sometimes there are gaps between those two worlds—the individual and the organizational culture and values. To bridge these gaps, to help an employee to adapt to corporate culture where necessary or—in worst case to stop cooperation—is a key management task on every level on the hierarchy concerning the management of the staff.

The "conscious mind" of an organization is represented by mission statements, visions and resulting corporate strategies and strategic objectives as essential elements for good alignment, focus and control of a company or organization. They officially try to convey the corporate culture. But real life is sometimes different. The "real life" of an organization—and thus also a part of the culture of an organization—is significantly influenced by the thought patterns and the discipline of thinking in the making of decisions as well as from role models, best-practice (as well as worst-practice) examples, leadership etc.

Starting from the mastery of comprehensive system thinking (see "soft infrastructure", Sect. 3.1.1)—optionally applying the system archetypes for understanding the dynamics and non-linear relationships in the mostly complex systems (learning and adaptive systems see Sect. 5.2.9)—the guiding principles from

Sect. 5.1.3 (the processes of cognition and decision) provide good guidance for the design and the training of the conscious mind and—indirectly also—the subconscious mind of an organization as well as their "care". Finally the way how an organization is managed and how that is perceived, is also formative for culture and the "conscious mind" of an organization.

> **GP32 Cultivate your organization's visions supported by system-thinking and "leading by example"!**
> Embed and anchor a consistent discipline of thinking in the organization. Common or similar mental models in teams and in senior management help to implement a jointly supported vision for the organization in terms of consistent objectives. Systems thinking and commitment to the vision and to the objectives influence the behavior of the employees and of the organization as such. In particular, the corresponding active approach of managers and experienced staff in the sense of "leading by example" is necessary in order to resist the tendency of complex systems, to counter all attempts to change their behavior ("policy resistance"). Team Learning—coupled with excellent leadership and trust—thus shape and develop the conscious mind, culture and the subconscious mind of an organization in its daily activities.

Below some cultural elements of an organization are summarized with a short explanation of these elements (see more detailed considerations on some of the issues mentioned in Sect. 5.1.3):

Discipline of thinking, mental models

This includes systems thinking, the discipline of structured thinking with a clear distinction between facts, findings, conclusions and recommendations. The underlying assumptions are made transparent, working hypotheses are formulated and questions derived in order to determine the relevant facts and findings. The training and the mastery of these methods and techniques result in an intuitively guided and better quality of thinking and reasoning.

Develop and live a jointly supported vision

The vision shared by as many employees as possible ideally is comprehensive and vivid as a hologram. The individuals and organizational units then ideally not only behave in accordance with these objectives in terms of "compliance", but they profess in the sense of "Commitment". In the highest form the vision creates a common purpose, which is pursued in partnership with each other.

Team Learning

Learning from and in teams is crucial for the conditioning and training of the conscious mind and thus the system 2 of the organization. It also conditions the subconscious mind of the organization, led by the shared vision and thought and argued in similar models of thinking.

Leading by Example, Trust

Nothing can damage so much in an organization as executives who do not exemplify the vision and strategy or even thwart them. For example, if the culture of absolute focus on the customers or beneficiary of the services of the organization is not exemplified by the executives, you will not be able to expect it from the staff. The same applies concerning an open approach to change. This does not mean that the sales manager must be the best salesman or a member of the board be a better financial manager, sales or operations manager. It would be absurd to demand that and it would only demotivate the staff in their core competency. It's about how the executives behave in difficult situations, in a crisis but also how they handle success —which attitude the executives communicates, how they consult with their staff, how they communicate their decisions, etc. Reinhard K. Sprenger says, "Leaders have followers". Voluntary followers are the indispensable counterpart of executives. The art is to bring the situation and the individuals to a fit (Sprenger 2013). To follow "voluntarily" means: The absence of coercion, to want in the sense of free will. But voluntariness also requires trust. "The mechanism of trust not only reduces the cost of cooperation, but finally allows collaboration. Collaboration without trust would be impossible or at least very expensive to materialize." says Sprenger. "In an uncertain world the need for trust and confidence is growing rapidly." Trust is reflected in the concrete behavior of individuals notably in case of conflict. Trust draws its strength from the collective favorability of confidence. Everyone is at the same time donor of confidence and trustee. Trust is a part of the culture of an organization and is located largely in the subconscious mind of the organization. Trust often is only visible in the intuitive behavior in "gut decisions" etc. (Thinking fast—System 1 by Kahneman).

The "conscious mind" of an organization with its decisions and actions taken truly consciously and reflected covers areas such as strategy, research and development, planning, marketing, accounting, etc. These are essential parts of the system 2—thinking slow—of an organization (following Kahneman).

Now a few comments on the topics of leadership, responsibility and motivation based on Sprenger. I can confirm most of it also from personal experience:

Leadership, responsibility, motivation

Leadership means self-management of the individuals, ideally to the point, where the employee does not need more guidance. Leadership and self-management need self-confidence. Will is the core of self-consciousness. Specifically will is the substantial experience of the individual's own identity—it is their "Me".

Self-management according to Sprenger means: to have my own motivational settings and to control them, to make decisions wisely and weighed, not having to rely on steering and extrinsic motivation. For self-management, self-responsibility is indispensable. To "lead" is the directing to "perform self-responsibility":

- Self-responsibility means to select—autonomous and voluntary action
- Self-responsibility means to want—initiative and committed action
- Self-responsibility means "to respond"—inventive and creative action.

Leaders who motivate for self-responsibility open the space for self-management. They thus enable self-organization and allow others to build their organizational processes and structures, as we shall see later.

The employees must feel that they are trusted. Leaders must avoid that the "will" of the employee becomes a "shall".

To lead—Management—Leadership:

These are three terms in everyday language that occur repeatedly synonymously in "management-English". Let me elaborate the concept of leadership, which is very commonly used, especially on the level of the conscious mind of an organization. But leadership also greatly affects the subconscious mind of an organization. Malik (2007a, b), one of the leading management thinkers of Europe, explained: "When differentiating between management and Leadership, you have to start from a possibly positive picture of management and ask from there what Leadership means beyond. There are many executives who are future-oriented, who have foresight, who are innovators and who comply with all the criteria of a positively understood leader; but as people they are much too modest to ever refer to themselves as a leader or to designate themselves as leaders. This would appear to them as arrogance. It is enough for them to be considered as good managers and leaders. I mean, that really good organizations can do well without that concept of leadership."

5.3.3 Thinking Organizations New—Process Organization

In 1994 Hammer/Champy had set a milestone in the management theory with the management classic "Business Reengineering" (Hammer and Champy 1994). In this concept they had placed the focus on process management. Thus they replaced the hype around the topic of Lean Management, which had dominated the second half of the 1980s and itself had already included many elements of process management. So in fact they expanded the concept of lean management and complemented it. (The term "lean management" is actually getting popular again in management literature, for example in "lean hospital management", trying to focus all aspects and especially the processes on the patient—also harnessing the benefits from digital transformation). Lean Process management is now already practiced in many organizations. If someone wants their organization to be certified—no matter what standard (ISO, EFQM etc.)—process management will always be a core topic.

Many certifying consultants even began to define process metrics and urged for measurements. This—if not implemented very focused and sparingly—soon turned out to lead to a lot of bureaucracy in a variety of processes. The relevant notified process managers soon graded themselves equally to line managers and urged for corresponding claims. As a result of a long learning curve, process management confined itself increasingly to a few critical management processes and to mission-critical core processes. The novelty subsided, and other management issues came to the fore, such as Change Management. With the definition and measurement of management processes we pursue the goal of continually and comprehensively improving the process. A comprehensive description of a process and its environment from my experience ideally includes the following elements:

- Process context
- Process goal
- Process results regarding structural quality, process quality and outcome quality
- Process description with triggering events, process steps and the closing events of a process
- Process indicators (where appropriate).

This comprehensive description can be easily applied to key management processes. With business core processes, the focus is more on the process steps and the flexibility for ad hoc changes. The value here often is to involve all stakeholders in a structured manner and to work on—mostly graphically—the description of the specific process and bring clarifications between different views—often based on exemplary reference processes. When the processes are running routinely, the process descriptions are rarely used, unless major change needs to be discussed. In processes rarely needed, a process-based view can help to remind the rules to be observed and to ensure the proper flow of information—for example in case of handling a crisis.

The described processes are—in an analogy view—something like the musical score for the various instruments and voices of an orchestra or a choir. With well-established orchestras, the conductor can be limited to very economical interventions in directing. One often wonders what would happen if the conductor would not be there....

The process descriptions are often an indirect reflection of the organizational structures and their complexity. The data structures, in particular the levels of aggregation and the associated key figures in management information systems form another picture of the organizational structure and its inherent complexity. The new in-memory computing technologies and visualization capabilities make some complexity become obsolete. Former technical restrictions and reasons for building aggregates in IT-systems for example indicators as order entry per month or revenue per month have become obsolete by enabling systems to calculate these figures on the fly from basic recorded events like order entry, accounting etc. This can also be used as an impetus for rethinking of the organizational structures. Another indicator for sometimes too complex organizational settings—next to extensive process descriptions and hardly comprehensive reporting structures in the

Management Information System—are too complex project agreements where in complex organizations an unusually large number of participants are sitting in project committees. Everyone has a bit of responsibility and wants to have a say— in the worst case some want to cause some trouble and "bring in" some well argued blockade in order to "take out" something for themself and their organizational unit.

Today the challenge is often to be faster, more flexible and more agile than the other. Through skillful parallelizations of process steps often substantial improvements can be achieved.

The producing and service providing business processes—often called core processes—usually represent the core transactions as an essential part of the current business models. It is essential to understand the innovation of processes as an essential part of business innovation. Traditionally, product innovation is mostly in the foreground. Although product innovations usually result in revenue; whether it is profitable is mostly decided through process innovation (see Suter and Weitlaner 2014). The fusion of product and service in the form of "hybrid service bundles" or "service-enabled products" emphasizes these aspects of "business innovation".

Oxford Economics (2011) thinks that the pure focus on conventional business transactions would mean a loss of knowledge for the organization. The exchange of knowledge takes place only if cooperation is embedded in the business process. The design of business processes so has to do quite a lot with knowledge transfer and knowledge discovery. There is a reason that—compared to previous IT technical implementations of business processes—now enhanced techniques are implemented in very close integration with the data, information and knowledge sources and also integrating social media techniques.

The user and responsible employee acts—loosely guided by the business process with sufficient possibilities to ad hoc insert an intermediate step or possibly look at the suggestion of the decision support system or other knowledge and information. They ideally have the goal in focus and hopefully the vision, task and responsibility in mind. The systems support and guide them in their activities— keeping the individual employee responsible for what he does.

GP33 Accept some loss of control in favor of self-regulating flexible process management!

Create a flexible process management environment that is well-connected with the company as a whole and the internal and external partners, customers and sources of knowledge. In this environment, responsible employees decide and act on time, comprehensively, quickly, flexibly and intuitively—in accordance with their task. Accept that this might entail some loss of control and some unpredictability (due to emergence). Evaluate the risk grossly and accept it unless it seems critical. Enable collaboration on a social as well as on a technical level.

Example industrial company—*new structured maintenance processes and responsibilities*
For **Martha, the managing director** of the **steel-manufacturing company**, it became increasingly clear that the analysis of sensor data was a dynamically developing technology in the rolling mill. Its benefit depended on the data fed by the mill operators—her company as well as other rolling mills with a similar profile. A "cross analysis" only was possible for a plant manufacturer. Although her maintenance engineers and managers were very skeptical about the project, Gabriel already recognized the impossibility of being able to hold this expertise alone in their company. In addition, other operational units such as the steelworks and heat treatment facilities were in similar situations. The pioneer in their company was the rolling mill. Martha now mandated **Gabriel** to define collaboration processes with the system supplier so that their company itself lost as little of its autonomy and sovereignty as possible, and yet was able to exploit the technological developments. The authority for operating—when to schedule a maintenance-related shutdown etc.—had to stay with their company in any case. A second objective was to link the remuneration of the partner on the technical availability of the rolling mill and the maintenance costs—at least partly.

5.3.4 Thinking Organizations New—Organizational Structure

The above discussed concepts and analysis of organizational structures and process management has shown that the organizational structure often has been and mostly still is a source of—sometimes unnecessary—complexity, and that the improved collaboration with internal and external partners as well as with customers and the access to internal and external knowledge sources are becoming increasingly urgent. This requires not only a flexible connection of all these elements in a secure network providing the best possible user experience and intuitive ease of use. But it also suggests new procedural and organizational approaches. Together with the new technologies already available and those already visible on the horizon that seems to require a "rethinking of the organizational structures". This is about structures of power and influence as well as income levels etc. Some experts believe that "social collaboration" and "cooperation" will be the new working paradigms, that have to be provided with suitable infrastructure as well as supported by a proper accordingly designed organization.

Amartya Sen, Nobel laureate in economics, has said: "Development is the process of expanding freedoms". It is also about the reorganization of responsibilities. Sprenger—overlooking today's organizational structures—speaks of commonly "organized irresponsibility" (Sprenger 2013).

Many hierarchical levels, as we often see today, allow more easily for the top hierarchical ranks to pay themselves high salaries. The step by step decrease towards the lower salaries of those "working below" is easier to argue when there are many hierarchical levels. This of course has an impact on the readiness to change and the resistance of the actually responsible structures in the upper echelons.

The greatest pressure for change may well be exerted by the markets. The worldwide rapidly changing markets are forcing traditionally oriented Western companies to increasingly move away from the strictly hierarchical decision-culture and to provide a market-like "organic" network structure—not unlike the Internet itself. It applies to large international corporations, which have to take advantage of their size on the one hand and still need to be agile. Successful international corporations increasingly no longer add local national companies together under one roof, but are placing corporate functions where they find favorable cost, skilled labor and raw materials and where they best can support the core business processes and management processes. Advances in information technology and in business and data analysis make it possible to track the performance and market developments timely and to control and intervene appropriately.

According to Dr. Sviokla of Price Waterhouse Coopers, such companies navigate through the above-described global integration and develop towards "edge-based organizations": In this type of organization, the leaders and their teams will be authorized and given high responsibility on the "edge of the organization"—at the external border towards the customer and to the market in order to act very autonomously. These "edge-based organizations" are characterized by the ability of self-organization, distributed decision making and high market adaptability (see Oxford Economic 2011). The overall organization needs in its structure a reasonably and well-designed "subconscious mind" to meet the vision and the objectives of the overall organization and to provide the appropriate framework for self-control.

The organization of a "Task Force" or a "Special Force" as temporary unit, where everyone has a clear awareness of the initial situation, the values, the vision and the goals of the organization, where everyone knows where to find the necessary skills and where everyone has the authorization to take measures or give appropriate power to decentralized units, enables the rapid implementation of targets.

President Obama had equipped his campaign organization, already in the 2008 election campaign, with all means for self-organization and dissemination of his messages through digital tools. By constantly updating the distributed and decentralized organization and by timely monitoring of the action with modern digital means, the front people were empowered to convene decentralized meetings, to launch email campaigns and messaging campaigns, to make surveys, etc. In previous electoral campaigns this usually had to be authorized centrally. For the central election management this "subconscious mind of the organization" worked excellently and they were able to bring key messages quickly and efficiently to the people. Mistakes were detected in time by central monitoring and "alerting" based on observation of—among others—the social media channels and thus could be countered quickly.

GP34 Make your organization flat, fluid and adaptive!
Let us think our organizations new and let us organize them as flat as possible. Let us cover the necessary technical and managerial requirements: flexible and agile process management coupled with good networking with internal and external partners, with customers and with all relevant sources of knowledge. This has to be supported by information management systems based on the operational transaction data—without hierarchically oriented aggregate structures that augment complexity and reduce adaptivity.

The appropriate coupling of the infrastructure for cooperation and collaboration thus enabling best practice processes and process integration are the key technical challenges. These require proper information management strategies and the right platforms based on established common standards.

GP35 Allow self organization and distribute responsibility clear and transparent!
There is a yearning for clarity in complex structures. Allow and enable collaboration and self-organization. The new flat organizations need employees who are willing and able to take responsibility and secondly they need leaders who motivate their employees for self-responsibility thus enabling self-motivation, self-management and self-organization.

The organizational forms of temporary or permanent working groups, of a task force etc. can be implemented as a deliberate measure against intuitively induced reaction patterns of inertia—controlled or triggered by the system 1(thinking fast) of the organization's subconscious mind. These working groups question and challenge the resistent interests of the organizational units and lead to more structured and conscious solution finding—using system 2 (thinking slow).

A similar integrative effect can be developed by smartly induced certification projects and the related audit processes—if they are placed in the sense of and with the honest purpose of organizational learning.

GP36 Enable flexible and temporary working groups to increase agility! Implement temporary and permanent working groups, task forces, etc. with clear tasks and implement regular audits to "keep the organization in motion". They should help to loosen steady states and inertia and to find innovative solutions. A carefully chosen diversity in the composition of the working groups prevent unconscious response patterns of "run-in" organizational units.

Modern organizations are huge networks of interconnected entities. To cope with the challenging tasks requires socially and technically well qualified employees who are able to take the responsibility for themselves and who do not easily slide into burnout.

The most important management task will be the strengthening of self-responsibility in the company. In future there will be fewer managers with increasing leadership spans. The leaders must therefore empower more and "make" less. These are good conditions in order to create the opportunity for self-responsibility. "Self responsibility is the prerequisite for commitment and thus for excellent performance" Sprenger says.

The most important management tasks are:

- The selection of personnel,
- Staff development (including the taking out of employees from the responsibility.) with training and education,
- Enabling and encouraging employees to take responsibility and to act independently and to refrain from delegating back
- The agreement on goals and
- The monitoring of processes (in particular the decision-making processes), of customer and partner relationships and of the achievement of goals.

Employees need to know when the manager is to be involved and when they—from the perspective of the manager—are to make decisions without consultation in the "subconscious mind of the organization". They have to judge themselves where to decide intuitively (System 1—thinking fast) and where to invest time for detailed analysis (System 2—thinking slow). A flat organizational structure impedes too much back delegation and attempted shifting away of responsibility. Leaders who tend to define their importance and power by taking as many decisions as possible downright challenge and generate back delegation and thus will not succeed in creating an efficient flat organization with high leadership spans. The value and importance of the terms "manager" and "employee" will in future move nearer to one another.

A "well-adjusted subconscious mind" of an organization allows a relatively effortless and efficient system-1-mode and thus a good and efficient organization. Permanent action in the system-2-mode would be too uncomfortable and inefficient.

Working groups, project groups, experienced staff and managers have the task to constantly pay attention to the good "conditioning of the subconscious mind" in the respective level of the organizational cascade by temporarily and consciously putting themselves in the System-2-mode and reflect their actions in the continuous effort for improvement.

A key task in the increasingly networked world is the management of collaborations. This requires to clarify the goals and tasks in the collaboration on the one hand and to actively shape the processes allowing creative self-organization on the other hand. Cooperation and self-organization lead to an "active working together" in the sense of collaboration—"working together on the same objectives".

A difficult process is the adjustment of remuneration and incentive structures. The upper levels of management fear a flattening of the organization, because it might of course affect their compensation. Therefore the biggest resistance to change can be expected from this group—especially when they are younger and far from retirement. Remuneration and incentive systems tend to overly emphasize the hierarchical component and the hierarchical responsibility. As these are sometimes excessively remunerated they create tension and stress fractures in the organization in the form of the exodus or the inner emigration of good and valuable employees. After the transitional problems when changing to a flat organization, this would improve significantly.

The outplacement of executives is a major challenge in organizational transition phases. The older experienced employees involved often have problems in a wider and flatter organization—with maybe less personal power. They probably have problems with former subordinates working on the same hierarchical level. It has to be remarked that the prototypical knowledge organizations—universities—are used to "non-vertical careers" and they are not necessarily used to retirement at the height of power. Often—for example—a professor serving for a period as dean occupies this higher hierarchical position and has more power. After his period in this office he steps back and serves as professor again—usually without suffering "mental harm" from stepping back. It is important to establish this culture in other organizations and in particular in companies too. This is supposed to be easier than now, as many companies are claiming to be a knowledge organization. It will be specifically easier when remuneration schemes of the organization are in balance and fit to the flatter organizational structure.

In Sweden—where the retirement age is much higher than in Austria or Germany—it is already common that older managers and employees in their last years of employment (sometimes in partial retirement) support—where appropriate —aspiring young employees or project managers or coach them regularly or in conflict situations or they serve as mediators etc. They can advise younger colleagues without patronizing them.

Example hospital group—*organizational structure flattened—difficult transition but ambitious goals met*

Paul, the CEO of the Hospital Group was—reviewing the past years— proud that he succeeded in eliminating a whole management level within the Group. He and his fellow board members now had a wider span of control; they had increased the pressure on the managers of his company to act independent and goal-oriented on the other hand. The motivation was high. The agreed goals with a variable salary component had indeed been a point of discussion in the Supervisory Board, but they had been well received by the majority of the managers. The current fiscal year looked good concerning the results—their ambitious budget goals seemed quite achievable. On the other hand, there had been some confusion and criticism concerning the coordination of and the collaboration within their flat organization—especially between the now very flat structured organizational units of the group management and the management teams in the various hospitals. Paul was convinced that the new flat organization and the not yet fully established new processes would soon be anchored in the subconscious mind of his company. In a few years he would know if the chosen path was too courageous or if it turned out to be a long-term success.

5.3.5 Integrated Management System as a Framework and "Operating System" of an Organization

One might think that the organizational structure and the position of the executives would lose importance with the introduction of a flat organization. The opposite is the case. People think in structures and find orientation in structures. They were and still are socialized in structures. Let us take an analogy from new medical technology: In tissue engineering—the cultivation of tissues—it is crucial that a structure is provided, so that the tissue or organ can grow and the cells can arrange themselves properly. They need the right expression of genes and the right environment in the form of a helping predefined structure.

All organizations operate in a specific environment with internal and external elements and "building blocks": organizational units, employees, locations, products, services etc. Employees and organizational units, community organizations, etc. have a conscious mind and a subconscious mind. Consciousness expresses itself in the deliberate efforts of an organization and is expressed in visions, goals, strategies, power structures, organizational structures etc.

Organizations usually already have a "conscious mind" when they are founded. Purpose, objectives and structure as well as basic partnership agreements are usually determined in the association statute etc.

The subconscious mind develops later, when the elements and building blocks of the organization interact, when infrastructures emerge, when processes are defined and modified etc. These developments can be influenced both directly and indirectly—as shown above in many facets—, but this influencing rarely has a direct and immediate effect. Usually the impact is a delayed one. Most is done consciously first and then gradually slips into the subconscious mind—it is very much how we learn to drive.

Conscious mind and subconscious mind of a specific organization must be in a certain harmony with each other and not in contradiction with each other. Otherwise, the organization will develop a "neurotic behavior" through to an "organizational psychosis". Then, however, the organization is most endangered in its sustainability and even in its survivability. It is therefore crucial, as goal conflicts are being resolved, how such decisions—which often are to clarify conflicting goals —are being taken.

The supporting structure along which the organization has to be developed— dedicated to continuous improvement—as a framework and a set of rules that I call "Integrated Management System". To paraphrase an analogy to information technology, one could also call it the "operating system of the organization".

Malik (2007a, b) has defined and described this term very comprehensively and brought to life, so that you can use it in the daily managerial practice. The integrated management system is presented in a kind of portfolio matrix in which, the business-related and employee-related dimension is shown on the x-axis and on the y-axis, the time horizon up to one year and the time horizon more than a year are shown. In this illustration, the elements of a management system with the interrelationship to each other are shown. This seems to be complex—like management is. Its complexity however can be easily comprehended—as with any cybernetic system. The elements of an integrated management system are specific to each organization in their characteristics and how they develop. In the practice of a specific organization they sometimes have different names, some elements are omitted, some added. The basic idea is compelling, and the original model of Malik is a valuable "two-dimensional checklist".

Below you find a list, where some elements of a management systems as parts in this "coordinate system"—from the top left (company-based, more than one year) down to the bottom right (employee-related, up to one year) are displayed:

- Corporate purpose
- Mission, vision, goals, strategies
- Organizational structure
- Some specific management systems concerning Human Resource Management
- Operational structure
- Work organization
- Results.
- Tasks
- Controlling

From this point of view or perspective you can gain a good overview of the functioning of an organization, and consciously define the framework for the respective organization. In this context the subconscious mind of the organization can then gradually develop—reasonably, purposeful and oriented towards the goals of the organization.

The individual management systems (such as information management system, quality management system, risk management system, internal control system, environmental management system, etc.) and regulations (such as compliance rules in the context of the "code of conduct", guidelines, etc.) are also part of a comprehensive and integrated management system. The management and its core processes, for example, processes such as "develop business objectives and strategies and their monitoring", the integrative processes of "economic planning, capacity planning, financial planning and controlling", "recruitment and development of key persons" and "managing contracts" etc., should be regulated in a coherent process management system and with a consistent method and mindset throughout the organization in relevant granularity—specific enough but not too detailed and bureaucratic. It is advisable to merge the rules in a management manual and to keep that clear, compact and up to date—as the "operating system" of the respective organization.

Management systems are an interface between the subconscious mind and the conscious mind of an organization. The instruments and the elements of an integrated management system as ICS (internal control system), risk management system, controlling, process management system etc. are the triggers that detect deviations, recognize the change and improvement, initiate the deliberate steps towards reform and recognize the need for involvement of the senior management etc.

Sometimes management systems induce–quite necessary—requirements that seem bureaucratic for the employees of the core business, which also could be seen as an expression of lack of confidence in the employees. This distrust sometimes even might be justified. Whether one or the other of these requirements of the management systems could be reduced and cut back—while simultaneously assessing the risks—is to be discussed by the respective management team according to each organization and the decisions have to be taken.

The future economic and social developments are increasingly demanding agile, flexible and adaptive organizations. Organizations full of distrust will never be truly agile, flexible and adaptive. It seems that it would be appropriate in this case for many organizations, "to think anew" and adjust the organization accordingly, as well as the management systems and especially the control systems. Confidence in the staff must remain the basic attitude of the organization. This confidence is backed by controlled procedures and spot checks and must not tilt in a distrust organization.

The management systems support—among others—the organizational structure. Structures shape the medium and long term behavior and—at least to some extent—also the attitude of the persons concerned. They also frame the understanding and communication of context. To that extent the management systems continue to influence the subconscious mind of an organizations.

5.3.6 Culture of Excellence

I do not intend to present the many different quality management systems in this book such as ISO 9000, Total Quality Management (TQM) or EFQM, which always have as one of their prime objectives to achieve a stage of excellence.

Instead, I would like at this point to present—based on Peter Senge (1990)—the learning process concerning a new or changed set of rules, values and assumptions aiming towards excellence in three stages (Senge is following in this point Diana Smith), all of which are to be gone through until you have reached the top level 1. These steps are—referring to the theme of this book—as follows:

- Level 3: Values and assumptions are adopted: People connect rules and link them together for their use, so that they can retrieve them under stress and in doubt. They have the rules, values and assumptions built into their own model of action, but they still refer to them in their own diction.
- Level 2: New rules of action are adopted: The old rules of action and behavior patterns are slowly disappearing because a deep cognitive insight in the new rules emerges. However, they still need the new diction to retrieve the patterns and combine them appropriately in each situation.
- Level 1: New cognitive and language skills emerge: The new rules of action are absorbed and internalized. You see the assumptions, activities and consequences more clearly and you express them differently, because the values and assumptions adopted at Level 3 become your second nature and become part of the "thinking fast" mode (System 1).

At level 1, the individual and the organization have reached the level of excellence. The new values, assumptions and rules are part of the subconscious mind of the organization and they guide the individual in the way he acts. But to go through this learning process takes some time and practice—with the individual as well as with the individuals interacting in the organization. It is wise to focus on the important and defining rules, values and assumptions. This should be possible if it is a strategically well-managed and aligned organization. It is much harder if the organization is just in a deep phase of reorientation.

Crises and strategic disruptions are something that is happening in organizations from time to time and they will always happen. There is something threatening about that, but it also offers real opportunities for innovation as well as for strengthening and ensuring sustainability. This takes us to the topic of resilience.

5.3.7 *Resilience*

Resilience describes the tolerance of a system to disturbances and thus describes the ability of a system to deal with changes (following Ungericht and Wiesner 2011). Systems have to compensate for internal or external disturbances thus coming to some sort of stability while maintaining their system integrity.

An illustrative example of resilience in the narrower sense is the ability of a tumbler toy: It can rise again from any position. The term resilience is interpreted differently in various fields. In case of a disaster, resilience is a collective term for dealing with problems of vulnerability of institutions or collectives.

Resilience management includes all measures with the aim to strengthen the capacity of an organization to compensate disturbing external influences. The resilience of an organization can be described by three characteristics or capabilities (following Ungericht and Wiesner 2011):

- Prevention: An ability for resistance to negative external influences is constructed as a precaution—it fosters the resistance of the system or the organization.
- Adaptation: A rapid adaptation to changed circumstances is possible—this requires self-regulation.
- Innovation: Emerging advantages of the changing environmental conditions are used economically—this requires the ability to detect opportunities as well as innovation management.

Diversity in education, social background and CV's of the acting persons is one of the essential characteristics for resilient organizations and structures, resilience-researchers say. The opposite of resilience was the fiasco of the 2008 financial crisis, which was caused by homogeneous, similarly minded circles and their synchronous, self-reinforcing behavior (following Schilcher 2013).

Basically, the resilience management can build upon and rely in its action on risk management and crisis management. Other effects, which contribute to increasing the resilience, are networking with stakeholders and a prioritization of learning. Thus, the resilience of an organization can be increased by a clear focus on the generation of knowledge and on increased diversification in the value chain.

It is crucial to prepare the organization for possible future negative events and not only repel negative events when they happen, but to transform them in a positive event if possible. A mastered crisis may well be something to be proud of in an organization. This can and should be formed into appropriate stories and be part of the self-image of the organization. This self-image and the self-confidence of the organization allows them to choose a preventive, systematic management system approach and to live it: The organization does not rely on luck or chance, but takes things into its own hands. The organization has a realistic image of its abilities (strengths and weaknesses) and thus also an appropriate self-confidence with which it appears and acts. Based on this self-image the organization will systematically seize the opportunities if they exist. Such stories of successful crisis management and the developed self-confidence of the organization are part of the subconscious mind of the organization.

Technical management of resilience requires an intelligent security management, which formulates their security issues in a "Security Policy" and builds the infrastructures accordingly. Particularly critical infrastructures have to be well protected. The allowable downtime and recovery time after physical destruction are to be defined, for example, for critical IT systems. Disaster scenarios are also to be practiced regularly. Regular training for example on the simulator or in other settings have to ensure that, for example, a safe and good medical care is possible even when routinely highly available decision support systems are down due to technical problems, blackouts etc.

Such technical failures or system breakdowns are not completely preventable and will probably occur in future. They are part of the uncertainty that is inherent in our lives and that is a characteristic of complex systems. So let us "embrace uncertainty" (following Nowotny 2015) and thus prepare for it by sharpening the required skills like staying resilient, detect opportunities and exploit them and shape the subconscious mind of our organizations. When uncertainty hits us, our organizations have to react appropriately and quick. Think back to our first metaphor— driving.

5.3.8 Managing Complexity and Embracing Uncertainty

Some aspects of how to manage complexity already have been considered in Chap. 4 of this book, where the model has been described and in the section about simulation. Uncertainty is on the one hand like a "child" of complexity with all its dynamics, non-linearities and interconnectedness and its tendency to contribute to uncertainty. An example: We build complex systems to control rolling mills etc. and everything seems to run smoothly. The uncertainties of the production process seem to be eliminated ....until the control system is hacked by cyber-criminals who try to blackmail the company. Uncertainty seemed removed at a first glance, but it has not disappeared—uncertainty just shifted to a higher level of complexity. New uncertainties as "children" of "mother" complexity seem to be born.

On the other hand uncertainty is like the "mother" of complexity. Uncertainty seems to induce complex systems in order to support the organization's attempt to reduce uncertainty and make it (seemingly) manageable—or better make us belief that. An example: We want to cope with the uncertainties of weather and build complex models and systems to simulate and forecast weather conditions so that— for example—the energy production of photovoltaic power plants can be better planned in order to efficiently distribute it into the power grid. These complex models and systems thus are the "children" of "mother" uncertainty—born from the wish to manage uncertainty.

As uncertainty frequently expresses itself with randomness, it is important to embrace the opportunities provided by randomness and fuzziness. Biological systems and evolution are at least built on the randomness of genomic and genetic processes. Thus uncertainty might be perceived as well as an opportunity for innovation and not only as a threat.

When designing and shaping organizations and their subconscious mind in particular you should follow the following guiding principle:

GP37 Reduce risk aversion, encourage to take responsibility, embrace uncertainty!
Facing high complexity, that you cannot reasonably further reduce, you should allow a sufficient degree of freedom in the different levels of your organizations and organizational units by reducing excessive (bureaucratic) control as well as overly risk aversion and allowing fast decisions on the respective level and by responsible employees—"thinking fast". Be courageous, motivate your employees for self responsibility, control the outcome to stay in the necessary authority of control—allow uncertainty, which is unavoidable anyway.

This will reduce complexity and the management effort for complexity—facing that uncertainty is inevitable anyway. In this context the right timing for intervening is essential. You have to know when it is time to act, to delay or to forgo. It is not the goal to be as fast as possible in the sense of "at once". It is the right time ("kairos") of action that is key. So "thinking slow" may be wise very often but it requires experience and patience and an excellent feeling for the business you are in. Remember the metaphor to the human subconscious mind and its influence on decisions—not only when you decide or act fast (system 1) but also when you take some time to evaluate and then decide (system 2). Your subconscious mind shapes and influences your decisions. The right timing (kairos) is not mainly a question of rules but a result of experience and feeling, something even promising algorithms from Artificial Intelligence won't provide for some decades.

I assume that AI will augment uncertainty because the reaction to events will be very similar every time the event occurs. The variety in responding and acting on certain events thus will reduce compared with humans taking their decisions in complex situations. Experienced und skilful humans will for a long time be more differentiated and be more sensible to the actual situation as a whole—they will be better in deciding and acting on complex situations than AI will be in near future and in a midterm view.—As we know from mutations in biology variety is key to evolution as well as variety and diversity are key to innovation and progress. (see Johnson 2010).

Let us look at a traffic situation to illustrate this comprehensively.

Example Traffic Situation:

On my way to work I daily cross an intersection of five streets with cars, long buses, many bicycles and many pedestrians due to the nearby university. All traffic lights had been removed, there are no (painted) crosswalks. This intersection is organized as a so called „shared space". When I drive there with my bicycle, I check if the drivers in the cars see me and I cross when I see from their body language, that they have noticed me. It is the same with pedestrians and other bikers. I never saw an accident or truly critical situations in years and there were hardly traffic congestions even in rush hour. Now imagine one or more self driving Google Cars try to enter this intersection with its numerous pedestrians and bikers simultaneously. My prediction: The self driving cars would not move and block all car and bus traffic because they would not dare to move into the intersection. Only pedestrians and bikers would move; the result would be a veritable traffic congestion. I presume this will stay so for some time in spite of all AI technology like prediction, deep learning etc. Self driving cars might succeed on highways but not in traffic situations like this.

Maybe you try to transform this example to complex decision scenarios in your organization and figure out what definitely will not be suitable for automation by AI in the organizations you are in.

5.3.9   Forming the Subconscious Mind of Organizations: How and How Quickly?

Obviously you have read this book up to this point and probably you have asked yourself two questions:

- In which situation is my organization? Are these issues relevant for me and my organization?
- If yes—how fast can the subconscious mind of my organization be designed or changed?—How do I get started?

First problem: How can I assess the initial situation? Is there any necessity for action in my organization or organizational unit? As a starter to this assessment try to answer the following questions for your organization (this list of questions is applicable as a basis for assessment for any kind of organization be it an SME or a large corporation or be it simply an organizational unit of an organization—surely not all questions are relevant for each kind of organization or OU):

- How do we take decisions at the various levels? Are we well prepared to take these decisions? Do we think and decide in a structured way? Do we act mostly intuitively? Are our decisions comprehensible? Are they well communicated? Are there mistakes and what do they affect?

- How does something ever come to the attention of the next level of decision making or the management? What triggers escalation and how does escalation to the upper echelons work? Is too much to be decided on the respective level and in the respective organizational unit. Is too much being delegated back or is too much being blocked from me as responsible manager? Do those who decide have the informations they need?
- Do we work really focused and directed by strategy?
- Do we train our teams and our employees about decisions and decision making process? Do we have a comprehensive methodology for preparing and taking decisions? Do we need one?
- Do we have an integrated management system? Do we really need one?
- Does something like organizational learning take place in our organization? Audits? Certifications? Effective internal control system? Effective quality circles or similar? Team Learning?
- Are there any decisions that are expected to be automated or supported by Artificial Intelligence in the future—where and at what level are they expected to be taken?
- Do we use our opportunities for innovation (product and process innovation, business model innovation)? Are we at least ready for it?
- Which new technologies and methods (see Table of Contents Chap. 3 and if necessary the comments thereto) could become relevant for my organization in the next five years?
- What opportunities for innovation do we have in the design and architecture of our infrastructure especially concerning ICT (use the table of contents as a checklist 5.2.2 to 5.2.12)?
- Is our information management and—if it already exists—our knowledge management well equipped for that future? Do we have the right platforms for these future scenarios? Is our infrastructure ready?
- Where can we remove complexity? Is the complex organizational structure interfering with our way to the future?
- Do we already have a culture of excellence?—Or at least signs of this?
- Are we resilient? Do we check this by considering scenarios of crisis?—Do we need a "stress test" for our organization? How resilient are we really?

Should you come to the conclusion that you and your organization or your organizational unit have a need for action in several points, it is perhaps worth to refer to the conceptual model of the subconscious mind of organizations and to the guiding principles described in this book as a framework. Having this in mind

- generate ideas for improvement and change,
- prioritize them,
- make action plans for the important and urging ones
- rearrange the priorities evaluating effort and impact concerning your business strategy and your goals (based on the action plans above) and

- then boldly execute them.

This will support the sustainable development of your organization.

Whether the new technologies will be very relevant to your organization or in case you see less relevance in this area: Put the relevant spheres of activity on your mental map of the subconscious mind of your organization and the adjacent conscious mind of your organization. On-going developments and changes in your organization or your business environment can probably be more easily incorporated into this conceptual model. Try a conceptual stress test with powerful scenarios of change—performing some sort of mental simulation.

If you notice from the assessment of your initial situation that there is some demand for action, you have generated ideas for action and you try to put the actions on a timeline, to balance resources and orchestrate the action plans, the second question arises: How fast can the subconscious mind of your organization be designed and be changed? I give you some key points to help you with your considerations:

Thinking fast: Let us consider first the "thinking fast" (System 1 by Kahneman) in the cascading levels of our organizational model: the higher you come in considering the cascade from the individual employee (with humans thinking fast (System 1) is a matter of fractions of a second) to the entire company (routinely taken decisions on this level can take up to several days; if they are automated by AI it is as fast as the human subconscious mind), the longer it takes and the harder it is to change behavior and to leave the more intuitive system 1 mode (controlled by the subconscious mind of your organization with all their rituals) and switch to the more strenuous and exhausting system-2 mode. The changeover already has to take place as a conscious action. Bad habits, rituals and response patterns of an organization unconsciously strike and thus often thwart the good intentions to act organizationally consciously and guided or directed by strategy. These strikes are often stronger than the good intentions of an active management. This is one of the most significant challenges for executives.

To make the subconscious mind work efficiently and to recognize when to switch to organizationally conscious decision making is a process of organizational development and of organizational learning. To improve and professionalize this switching from "thinking fast" to "thinking slow" means to identify and dump "bad habits" and rituals that do not serve the organization's mission. As we all know, a change in behavior can be tough learning and sometimes requires a change of attitude on the personal level. It is similar with the change of organizational behavior. It can be a lengthy process. Keep in mind the three steps towards a culture of excellence (Sect. 5.3.6).

But it is also a question of the design of the infrastructure, the implementation of new IT platforms and new technologies and therefore a question of available resources for the investments necessary. Finally, it is also important to consider the aspects and costs of staff development and training programs including elearning, microlearning and other methods to influence the skills of employees.

Thinking slow: When we look at organizationally "thinking slow" (system 2 following Kahneman) we

- first have to make sure that we have consciously left the world of organizational rituals and habits—of our "automated organizational behavior" on the respective organizational level. So the question should be: Is it right, that we decide this here and now, is it right that we have to intervene in what is happening on the level below, in the organizational unit or in the project?
- If yes: Then traditional tools of structured, collaborative decision making get into action as they are part of our organizational culture and our business.
- If no or if you detected wrong decisions from your controlling perspective: do the same as with yes, but additionally consider to give clear feedback to those who have invoked or alarmed you or have delegated up to you. In case it was AI-driven decision automata or bad decision support given from a system: have them changed or disabled.

Dear reader, the contents of the preceding sections of this chapter on the design of an organization and their subconscious mind may serve as a reminder and checklist; the content and the guiding principles may assist you in the implementation in your organization. The illustrative exemplary scenarios may stimulate your imagination. But....

How to get started?

It requires a boost—an initial impulse—to ignite the learning and change process and get it going. A change in the management team, a profound strategy development process, changes in the organizational structure, outsourcing measures, significant changes in the ICT systems, as well as corporate crises can provide such impetus.

Those in charge should be a priori aware, that they will thus influence consciously or unconsciously the subconscious mind of their organization. Therefore, they should take the opportunity to initiate a deliberate design process of the subconscious mind of their organization (and its interfaces to the conscious mind of the organization)—with the considerations following the presented framework and thought model and the presented guiding principles.

So the initial impulse should be well prepared and accordingly be shaped unless it happens like an unforeseen crisis. An inventory of the initial situation (as described in the bullet points at the beginning of this section) has to be made in order to find, inspire and define areas of actions to be taken. Mind to look for Quick Wins. Usually they have to be consciously sought and implemented. When implemented successfully they have to be fed back to the organization accordingly as success story and motivator. No one will start a strategic project, named "We build our subconscious mind of our organization anew". There are more subtle approaches needed.

The focus and the rationale always have to be the customers, the products and services as well as the employees and the stakeholders. The concept and the idea of designing the subconscious mind of your organization should always serve as a kind of hidden agenda in the background, in the selection of areas of action, in the systems to be renovated and redesigned, in the priorities chosen and it will affect investment decisions, the decisions for projects to get started and it will shape the reasoning of these decisions.

As with any cultural change it requires a lot of "fitness" and consistency to make the changes a lasting effect—a learning process precisely. It requires leadership on the respective level of management. And all "players" on the respective levels, who lead, manage and perform their changes, projects and implementations (partly simultaneously) should have the same idea in mind. They have to be aligned in a strong management team with clear visions, so they can act autonomously and though synchronized in regular meetings for controlling and aligning. It would be too slow to work one after another for the complexity and connectedness would be too hard to manage. And it would be too slow.

The integrated management system and the changes in the management system and in the infrastructure of the organization should be thought through and prepared as early as possible, although the elaboration of details can be part of the implementation and the common organizational learning process. You should—according to the principles of good change management—"make the people concerned to become active parts of the change". So think of the following guiding principles when starting to actively form and shape the subconscious mind of organizations.

> **GP38 Scan your organization for already existing "subconscious elements"!**
> Analyse the initial situation of the organization with the concept and the metaphor of the subconscious mind as well as the conscious mind of organizations in your minds. Try to find scenarios regarding the relevance and possible application of new technologies and define the relevant areas of action. Always keep the aspects of organizational and cultural changes in mind.

> **GP39 Find initial sparks and quick wins to get the shaping of your organization's subconscious mind started.**
> Use enterprise-wide significant and visible changes as initial spark and impulse to initiate the transformation process of the subconscious mind of your organization. Align the integrated management system accordingly and early. Implement the right platforms and technologies as infrastructures. A good initial impulse and some quick wins with first rapid and visible success provide motivation for the transformation of the subconscious mind for the organization. The sustainable implementation requires consistency, patience and vigilance—an ongoing organizational learning and staff development process.

To summarize and conclude, I would like to give the following guiding principle and scenarios on your way:

GP40 Establish an integrated management system to provide coherence and resilience in your digital transformation!
Organize the respective organization as flat and agile as possible. Develop an integrated management system as a framework and "operating system" that gives your organization stability in its flat and agile structure. Support the development of your organization towards excellence by incorporating parts of the management system in the routine of the daily work and let it thus flow into the subconscious mind of your organization. Provide the necessary coherence. Do not forget to secure the resilience of your organization.

Example hospital group—_satisfactory working organization, good KPIs, agile and resilient_
Paul, the CEO of the hospital group, could now be satisfied, looking back on four years in the new flat organizational structure, which resulted for him in a very wide span of control. Thanks to the integrated management system, which the board had developed with its managers in the first two and a half years and which had been documented in a comprehensive management manual. The flat organization soon was working in a satisfactory manner. The annual agreements on business targets with the managers and the business plan conferences with the management teams of the hospitals were a lot of work. These conferences provided a strong bonding experience under the umbrella of the corporate strategy and they helped to refine the strategy. The ambitious budget targets had been achieved without loss of performance. Even compared to other hospital operators, the indicators of his group developed excellently. Negative developments could be identified early and countermeasures could be set. Even difficult structural changes could be implemented quickly and without too many problems. The self-confidence had grown and the subconscious mind of the organization was focused on the structural challenges providing healthcare as well as the welfare and safety of patients—even in tight budgetary constraints. Paul wondered, however, whether they would survive crises due to lack of resources. Was his company resilient too? From the perspective of Paul they were well prepared. When an even tighter budgetary framework would be given, the owner and the politics would just have to allow structural changes. In change management their flat organization had already proven to be successful in recent years.

Example industrial company: *profitable and flexible steel producer with strong maintenance partners, variable costs and good governance*
Martha, the managing director of the special steel producing company, now after three years, took stock of the new maintenance strategy, which—after the pilot tests in the rolling mill—had been extended on the whole company. The maintenance partners—usually the largest equipment supplier in the respective operational unit—had taken over essential tasks of system monitoring. They used the data of her company as well as the data of their other customers with Big Data technologies in order to safeguard the availability and reliability of the production equipment by early recognizing the looming repair and replacement needs before an unplanned disruption occured. The respective spare and repair measure where then taken over by the respective partner in close coordination with the plant supervisors in the operational units, the remaining small but powerful maintenance team and the managers. Gabriel, the project manager of the pilot project in the rolling mill, had become head of the company maintenance. The payment to the partners included a low flat rate and premiums for high plant availability and for high productivity. Martha now saw a much stronger effort by the plant suppliers to keep the machinery well sustained and "in good shape". Of course their partners had better knowledge about the plants and its machinery than the maintenance engineers in the manufacturing companies, especially since everything was "peppered" with electronics and sensors. At times of declining order intake the premiums were not fully paid because the high availability could not be turned into productivity. So Martha had made the costs more variable as well and could better follow the business cycles. But the maintenance partners also could live with it. In good years they regularly had additional income from the premiums. The coordination with the production planners worked out great. The previously customary weeklong maintenance-related shutdowns were reduced to one week maximum per operational unit per year. The production stops were perfectly matched and no longer led to these annoying and expensive inventory peaks in the summer months, when these previously two to three-week "plant summer holidays" were used for maintenance of all equipments simultaneously. The initial difficulties of the new organization had been manageable. Now—in the third year—all this had become part of their flesh and blood, part of their routine—and part of the subconscious mind of their organization. Martha was proud to have found a new form of cooperation with the partners who brought benefit for everyone. Thus her company had achieved a decisive improvement in its flexibility and its productivity. They now were the most profitable business unit in their Group—despite the high wages and low working hours in Europe.

One initial spark for "thinking your organization or your organizational unit new" is the digital disruption that is changing many businesses and that is leading to the digital transformation of the organizations involved. In the worst case organizations fail to adopt and go bankrupt or disappear from the market or are replaced by organizations better suited for the new times. Organizations to survive undergo a digital transformation—either in an active management approach—maybe using the metaphor of the subconscious mind of organizations and this framework as a tool and as a guiding idea when thinking their organization new. Or they go the passive way and let it happen, reacting on upcoming challenges in the best possible way and as fast as possible. This might not be sustainable in the end. An aggravating factor is the increasing emergence of Artificial Intelligence in the organizations which accelerates the change and makes the challenges even harder and more disturbing for the staff of an organization.

How to meet these challenges of ever closer collaboration between man and machine (AI)? In the following chapter we discuss the digital transformation as given meta-process in society and business and develop the metaphor of hybrid intelligences becoming elements of future organizations. How to deal with Artificial Intelligence entering our organizations and changing the conscious mind as well as the subconscious mind of our organizations by changing the perception, the cognition, the decision making and the executing of decisions in our organizations?

Literature

Arbesman S (2016) Overcomplicated: technology at the limits of comprehension. Penguin Random House, New York

Business Dictionary (2017) Definition of corporate culture http://www.businessdictionary.com/definition/organizational-culture.html provided by webfinance Inc. 17.4.2017

Csikszentmihalyi M (2010) Flow: Das Geheimnis des Glücks. Klett Cotta Verlag

Frackowiack R (2014) Die Objektivierung des Menschen. European Forum Alpbach, 18 August 2014

Gibb BC (2011) The emergence of emergence. Nat Chem 3(1)

Gleick J (1987) Chaos—making a new science. Penguin Books, London

Hammer M, Champy J, Künzel P(1994). Business reengineering: die Radikalkur für das Unternehmen, Campus Frankfurt

Harrasser K (2013) Körper 2.0 – über die technische Erweiterbarkeit des Menschen. transcript Verlag, Bielefeld

Himpsl F (2014) Sicherheitsspezialisten: Ganz schnell geklaut, DIE ZEIT N° 16/2014, http://www.zeit.de/2014/16/industriespionage-jobs-ingenieure/seite-2

Johnson S (2010) Where good ideas come from—the natural history of innovation. Penguin. ISBN-10: 1594485380

Kahneman D (2011) Thinking fast and slow. Farrar, Straus and Giroux

Kampits P (2014) Klüger, besser, schöner...- Ist der Mensch optimierbar? Vienna Lectures European Forum Alpbach, 19 August 2014

Kindermann S, von Weizsäcker RK (2010) Der Königsplan – Strategien für ihren Erfolg; Rowohlt Verlag

Kurzweil R (2005) The Singularity is near. Viking

Malik F (2007) Management – Das A und O des Handwerks. Campus-Verlag

Malik F (2007) Richtig Denken – Wirksam Managen - Mit klarer Sprache besser führen. Campus Verlag

Mayer B (2014) Mensch 2.0 - Roboter und Androiden. Wie mit Hilfe von Bionik künstliche Wesen entstehen; Dokumentation, Servus TV 20.8.14

Mayer-Schönberger V, Cukier K (2013) Cukier Kenneth; Big Data – Die Revolution, die unser Leben verändern wird. Redline Verlag

Modha D (2016) Introducing a Brain-inspired Computer—TrueNorth's neurons to revolutionize system architecture, http://www.research.ibm.com/articles/brain-chip.shtml

Müller T (2014) Arbeitsplatzsicherheit, Vortrag in Graz, ACP-Forum, Oktober 2014

Neuweg GH (1999) Könnerschaft und implizites Wissen: zur lehr-lerntheoretischen Bedeutung der Erkenntnis-und Wissenstheorie Michael Polanyis, vol 311. Waxmann Verlag

Niessner R (2014) Einsatzgebiete der Simulationsmethoden, WINGbusiness 1/2014

Nowotny E (2015) The Cunning of uncertainty. Wiley, London

Oxford Economics (2011) The new digital economy—how it will transform business, White paper from a research program sponsored by AT&T, Cisco, Citi, PwC, SAP

Schmitt (2014) So leben wir in 5 Jahren. Die Zeit 36/2014

Schilcher B (2013) Vortrag zu Resilienz auf der LSZ CIO-Konferenz, Loipersdorf, Österreich 17 November 2013

Senge P (1990) The fifth discipline, the art and practice of the learning organisation. Doubleday currency

Spitzer M (2012) Digitale Demenz – wie wir uns und unsere Kinder um den Verstand bringen. Droemer Verlag

Sprenger RK (2013) An der Freiheit des anderen kommt keiner vorbei. Campus Verlag, Frankfurt/New York

Springer Gabler Verlag (Hrsg) (2011) Gabler Wirtschaftslexikon, Stichwort: Tacit Knowledge, online im Internet: http://wirtschaftslexikon.gabler.de/Archiv/147159/tacit-knowledge-v4.html

Stephan A (1999) Emergenz: Von der Unvorhersagbarkeit zur Selbstorganisation, vol 2. Dresden University Press

Stephan A (1999) Varieties of Emergentism. Evol Cogn 5(1)

Suter A, Weitlaner D (2014) Innovation von Organisationen und Prozessen—Grazer Ansatz für Organisations- und Prozessgestaltung. WINGbusiness 3/2014, Graz

Topol E (2014) On the medical geographic information system, clinical informatics news, 31.7.2014, http://www.clinicalinformaticsnews.com/cln/2014/7/31/eric-topol-medical-geographic-information-system.html

Ungericht B, Wiesner M (2011) Organisation & Change Management—Resilienz: Zur Widerstandskraft von Individuen und Organisationen. Zeitschrift Fuhrung und Organisation 80(3)

Wipfler C, Müller S, Vorbach WA (2014) Marko Hybride Leistungsbündel – Wenn Produkt und Service verschmelzen. WINGbusiness 3/2014, Graz

Chapter 6
Will Organizations Emerge as "Hybrid Intelligences" from the Digital Transformation?

Abstract Considering that new technologies and especially Artificial Intelligence offer new possibilities for strong bonding within the organization, but also in the interaction with others, this brings us to the question as to whether organizations can develop and incorporate something like "hybrid intelligence". We have three metaphorical perspectives that help us to gain a comprehensive view on what is lying ahead of us: • The **digital transformation** as a societal meta-development—as such directly impacting organizations, • the **subconscious mind of organizations** for the question of structural and process-oriented aspects of organizations as well as for methodological and technological questions of decision support and automation and • the organization that has incorporated **hybrid intelligences**. This "Big Picture" of the future follows an obviously inevitable but influenceable development of the cultural evolution of mankind—might that be considered positively with optimism, might it be considered with caution, might it be considered extremely critical. These perspectives and the necessary rules regulation, legislation and ethical considerations may help us as a guide and as reference points in this phase of the socio-cultural and social paradigm shift we are in. Together with the guiding principles for shaping the subconscious mind of organizations from the previous chapter the section titles of this chapter about emerging and actively shaped "hybrid intelligencies" may serve as additional guiding principles for building sustainable organizations.

Keywords Digital transformation · Hybrid intelligence · Decision support · Rules and ethics for Artificial Intelligence · Decision making · Man and machine

Are these organizations that we have previously discussed, with their abilities to automate actions, controlling and coordination, already some sort of manifestation of Artificial Intelligence? Let us consider again a summary of the characteristics of Artificial Intelligence (AI) and let us reflect on rethinking organizations from this point of view:

- Language is one of the essential means of expression of our mental processes. The representation of language in the computer sciences has significantly shaped the development of information and communication technologies. The Turing test as a criterion for the detection of Artificial Intelligence is an expression of this decisive role of language: The observer has to determine who the computer and who the human is by asking questions. If the observer is not able to reliably determine, who the computer is, the Turing test is passed.

- Language constructs also shape our thinking and are therefore part of our conscious mind, but also of our subconscious mind. The mostly philosophically discussed "mind-body problem" deals with the question whether living things possess something like a soul, a mind or a free will.

This question will in future have to be discussed for AI too. The "Group selection Theory" says that in evolution characteristics have prevailed, which have brought benefits for a group, even though they have brought disadvantages for the individual. From this point of view we can imagine AIs are becoming quite powerful "groups".

This brings us to the question as to whether organizations can develop something like "hybrid intelligence", considering that new technologies and especially Artificial Intelligence offer new possibilities for strong bonding within the organization, but also in the interaction with others.

For this purpose, we first try to find a compact answer to the question how the digital transformation as a whole is about to change the behavior of organizations as well as their purpose and mission. The specific developments in detail—the technologically enabled new ways of perception, of cognition, of weighing decision options, of finally deciding for one option and of taking action—have already been considered extensively not only from the perspective of the individual, but also from the perspective of an entire organization or an organizational unit. This resulted in the model of the subconscious mind of organizations and the metaphorical considerations derived from psychology and cognitive science in the chapters above.

Next we will consider the cooperation between man and machine—the "merging" of man and machine. This we will do in the context of an organization as a future form of a "hybrid intelligence" or at least of an organization with hybrid intelligences incorporated in her structure and her processes. This will lead us in the final section of this chapter to discuss rules and ethical aspects of such organizations with characteristics of hybrid intelligence.

6.1 How Is the Digital Transformation About to Disrupt the Way Organizations Work?

Digital transformation is the changes associated with the application of digital technology in all aspects of human society (Stolterman and Fors 2004). In a narrower sense, "digital transformations" may refer to the concept of "going paperless". Digital transformation affects both individual businesses and whole segments of the society, such as government, mass communications, art, medicine and science.

How to transform? According to CAP Gemini, in a study with MIT (Westerman et al. 2011), for an organization an "effective digital transformation program" is one that addresses

- "The What": the intensity of digital initiatives within a corporation
- "The How": the ability of a company to master transformational change to deliver results.

How far is my organization in the digital transformation? The report by MIT Center for Digital Business and Deloitte in 2015 (Kane 2015) found that "maturing digital businesses are focused on integrating digital technologies, such as social, mobile, analytics and cloud, in their efforts to transform how their businesses work. Less-mature digital businesses are focused on solving discrete business problems with individual digital technologies." Technologies such as the Internet of Things (IoT) and cryptocurrencies like bitcoin are already becoming an integral part of digital transformation.

Where does the journey lead us? With the rapid growth of the Internet of Things (IoT), tens of billions of sensor devices are projected to connect in the next decade. These connected sensor devices will automate processes across a broad range of economic sectors, from industrial plants to healthcare management, delivering productivity gains and hopefully quality-of-life improvements, Raja Jurdak postulated in 2016 (Jurdak 2016). The core of these sensor devices that will be deployed across this broad range of applications is largely the same, featuring a microprocessor, memory and a wired or wireless communication interface to the internet, along with a battery or other energy source. Each application and IoT device will bring its own unique context, such as its location, the conditions of the surrounding environment and the behavior of people in the area. Individual devices will observe and adapt to their unique contexts.

With AI, these devices can evolve their behavior in response to changing contexts. Just like how living beings optimize their behavior to their surroundings, even smaller IoT devices around us can run AI machines that evolve their software over time (Jurdak 2016).

Thus the business world is rapidly digitizing, breaking down industry barriers and creating new opportunities while destroying long-successful business models. Though this process called digital disruption with its sweeping technology-enabled change often takes longer than we initially expect, history shows that the impact of such change can be greater than we ever imagined. Think of steam engines, cars, airplanes, TVs, telephones and, most recently, mobile phones and e-anything. Can we today imagine a life without ubiquitous electricity like it was 100 years ago? Will people in 100 years from now be able to imagine a world without ubiquitous "cognification" as it is now beginning to emerge?

According to MIT Sloan and Deloitte in their 2014 Digital Leadership Executive Survey (Kane 2015) organizations across the board are using digital technologies to improve efficiency and the customer experience, but higher-maturity organizations differentiate themselves by using digital technologies to transform their business, allowing them to move ahead of the competition. "Senior leadership must really understand the power of digital technologies" they say.

Douglas Engelbart, an American engineer and inventor, and an early computer and Internet pioneer said: "The digital revolution is far more significant than the invention of writing or even of printing". So Digital is transforming business models across all industries and managing complexity will become one key challenge for the digital future.

6.1.1 Business Model Innovation Is Key for Creating Value

So let us take a look at the core of each organization—its business model.

Following Amit and Zott (2012) we can define a company's business model as a system of interconnected and interdependent activities that determine the way the company "does business" with its customers, partners and vendors. In other words, a business model is a bundle of specific activities—an "activity system"—conducted to satisfy the perceived needs of the market, along with the specification of which parties (a company or its partners) conduct which activities, and how these activities are linked to each other.

Thus more and more organizations now are turning toward business model innovation as an alternative or complement to product or process innovation, because in the operations area, much of the innovations and cost savings that could be achieved have already been achieved. It's not enough to make a difference on product quality or delivery readiness or production scale. It's important to innovate in areas where competitors do not yet act—in case your organization is in a competitive business.

Content, structure and governance are the three design elements that characterize a company's business model. Amit and Zott suggest that managers ask themselves the following six key questions as they consider business model innovation:

- What perceived needs can be satisfied through the new model design?
- What novel activities are needed to satisfy these perceived needs? (business model *content* innovation)
- How could the required activities be linked to each other in novel ways? (business model *structure* innovation)
- Who should perform each of the activities that are part of the business model? Should it be the company? A partner? The customer? What novel governance arrangements   could   enable   this   structure?   (business model *governance* innovation)
- How is value created through the novel business model for each of the participants?
- What revenue model fits with the company's business model to appropriate part of the total value it helps create?

You might now miss the digital transformation itself in these questions. Try to answer these questions considering what you read about the new technologies in the chapters above—and combine it with disruptive technologies in other domains like nanotechnologies affecting material science, 3D printing of multiple novel and established materials in new forms and shapes (bionics) as well as synthetic biology. You will surely find aspects that are relevant for your organization and you will find potentials that could disrupt your established business models (see some examples in the scenarios from the hospital group, the special steel company and the retailer in the previous chapters).

Aside the business model considerations—isn't it also a question of attitude and culture if you want to succeed with the digital transformation of your organization?

6.1.2 Culture as Prerequisite to Master the Digital Transformation

Let us start our considerations about the role of culture in the digital transformation having in mind that responsiveness to the market as well as outperformance in the business require an excellent strategy following Peter Drucker who once stated: "Culture Eats Strategy for Breakfast", thus expressing the dominion of culture, when it comes to implement strategies.

A 2013 Booz & Company study, "Culture's Role in Enabling Organizational Change" analyzes the results of a survey of 2200 executives, managers, and employees from a broad range of companies across the world. This research sheds light on current perceptions of organizational culture: "Your culture is the collection of words, actions, and perceptions that explains and supports what is really valued by the system".

That means it's complicated. There are four factors in this definition that create organizational culture: language, actions, perceptions, and tangibles. All people in an organization activate the first three. Culture is a complex combination of what they say, of their behaviors, and the underlying assumptions and ideas behind it. Organizational cultures can be viewed as living ecosystems of purpose, values, and behaviors (Aguirre et al. 2013).

Additionally, the "tangibles" have to be factored in. Tangibles are the *non-human* parts of culture. Tangible things like dress code, office location, office decor, office layout, etc. "These also reveal what is valued, so they have to be included in the assessment of what organizational culture of a specific organization truly is", Aguirre et al. remark.

When are organizations finally digitally maturing? Based on the Digital Business Global Executive Study 2016 Deloitte (2016) comes to the conclusion: A culture conducive to digital transformation is a hallmark of maturing companies. These organizations have a strong propensity to encourage risk taking, foster innovation and develop collaborative work environments. Culture needs to support collaboration and creativity. Digitally maturing organizations are considerably less risk averse than their peers. More than half of respondents from less digitally mature companies see their organization's fear of risk as a major short-coming.

So every organization has to question itself: Does my organization have a digital strategy that goes beyond implementing technologies? Does my organization's culture foster digital initiatives? Digital strategies at maturing organizations go beyond the technologies themselves. They target improvements in innovation, decision making and, ultimately, transforming how the business works.

Deloitte concludes "that many organizations will have to change their cultural mindsets to increase collaboration and encourage risk taking". This obviously has to become a part of their subconscious mind. "Business leaders should also address whether different digital technologies or approaches can help bring about that change. They must also understand what aspects of the current culture could spur greater digital transformation progress" Deloitte remarks.

Whether you are successful in your considerations about the future not only depends on understanding the relevant technologies and innovations, asking the right questions concerning business models, establish a culture of innovation etc. The starting point of your thinking and your considerations might make the difference.

"Every successful large-scale organizational change that we have seen has—as a part of it—a *change vision*", writes Peters (2015): And what that means is to have an image of what we *are going to* look like after we have made the changes. And if we look like that, our organization is going to be able to take advantage of the great opportunities that are a function of changes that are happening in this technology-enabled and hyper-connected world.

Benn Konsynski (Emory University), interviewed by Kane (2015) insists that the very concept of "change is changing." He explains: "A part of that means I start to work backward, not forward—the future is best seen with a running start. I'll go backward to better see the future possibilities—standing starts are poor means of seeing the future. … The constant caution, to me, is to make sure that we are not merely speeding up the mess."

6.1.3 Narratives as Tool for Strategic Management in the Digital Transformation

Now let us go beyond aspects of organizational culture as a foundation and basic prerequisite: What about strategic management in this era of digital transformation and—sometimes—disruption? A lot already has been said before. In turbulent markets, it can be hard for established companies to choose new strategic directions. One of the great challenges for organizations in the current economy is making strategy under the uncertainties posed by turbulent environments, intensified competition, emerging technologies, shifting customer tastes and regulatory change. Executives often know they must break with the status quo, but there are few signposts indicating the best way forward (Kaplan and Orlikowski 2014). How to find them? A core assumption in much of strategic management research is that more accurate forecasts of future competitive actions or the future value of certain business capabilities will lead to strategic success. They try to rely on Big Data and analytics. They try to embrace analytics and the use of data in decision making and processes.

But for many organizations, it is not the ability to process and manage large data volumes that is driving successful Big Data outcomes. Rather, it is the ability to integrate more *sources* of data than ever before—new data, old data, big data, small data, structured data, unstructured data, social media data, behavioral data, and legacy data.

This is known as the "variety challenge," and has emerged as the top data priority for mainstream companies, according to the fourth annual Big Data Executive Survey (Bean 2016). In the world of the Fortune 1000, they are seeing that variety trumps volume and velocity when it comes to Big Data success. Analytics seems to be pervasive, affecting practically every business. But obtaining competitive advantage from it is more difficult. Advantage is not achieved just by starting an analytics program, or by collecting more data, or by talking about it. As analytics diffuses to an increasing number of businesses, it has become table stakes. It's necessary to compete, but it's not sufficient to ensure to win.

"With a decline in competitive advantage, it might be natural to assume that we'd see widespread disillusionment with analytics", says Ransbotham (2016). Executives have long been exhorted to conduct analyses of internal and external environments and to construct scenarios of the future. However, seeing strategy in this way has some serious weaknesses. It assumes that accuracy can be achieved through rigorous analysis and conscientious efforts to overcome individual biases in perception. It also assumes that the process will be relatively frictionless and primarily analytical. There is an important tension at work here. Because the future is essentially unknowable, leaders must rely on the past for information and insight. Moreover, given that the future is unknown, effective projections of the future must be connected to resonant understandings of the present and past.

To develop a new strategic direction for a company during a time of change, managers need to create a strategic narrative that links the company's past, present and future. Crafting a strategic narrative involves reimagining future possibilities, rethinking the company's past and reevaluating present concerns. Thus by rethinking the past and present and reimagining the future, managers can construct strategic narratives that enable innovation. Kaplan and Orlikowski (2014) observed that, the more intensively managers reimagined the future, rethought the past, and reconsidered present concerns, the more their projects produced strategies that represented radical departures for the organization. It is not that technologies a priori represented greater or lesser change, or that new technologies forced people in the organization to engage more intensively, it is in constructing such strategic narratives, which makes digital transformation comprehensible for the specific organization and its staff members—that is digital leadership.

To illustrate this let us imagine a manufacturing company:

Example
Industrial company—*new business model with 3D printing of special steel*
forms in small lot sizes near the costumer
"The personal factory is here—they are going to push us out of the market"
Martha, the CEO of the special steel producer heard it over and over again in
the management meetings. Digital technologies have been knocking down
barriers to entry in tech and digital media—and now, they were witnessing a
similar effect for the makers of physical stuff. New low-cost, digitally driven
manufacturing devices are enabling people to do tasks that were traditionally
only possible in expensive facilities and factories—like her's.

Crowdfunding platforms were fueling the entry of these low-cost machines, like the Makerarm, which raised nearly half a million dollars on
Kickstarter, and is the first complete digital fabrication robotic arm that
mounts to a desktop (Berman 2016). Martha already read a lot about 3D
printing as well as additive manufacturing and its impact in different markets
and she had visited conferences, searching for strategic answers. She had
learned, that not only the access to new technologies was allowing tinkerers
and inventors to turn their garage into a factory. Startups in manufacturing
had triggered a revolution in how products are created—and massively
accelerated the timeline from idea to market. Product innovation has been
turned upside down, startups were flipping traditional manufacturing principles on their head. "So what is left to us as established but traditional manufacturing company?", Martha as CEO asked her fellow board members.
After some discussions and support from consulting firms, they came up with
a narrative for their future:

"We with our expertise, with our excellent brand and with our market
access will combine these forces. Rather than potentially wasting millions of
dollars in market research to create "the perfect" special steel products, we are
letting our customers play a key role in the actual product development early
on and incorporate this in our brand management. Let us specify—with our
experience in material science - new qualities for raw materials best suited for
3D printing of special steel components. Let us try it ourselves in our R&D
department. We are able to combine economic production in lot sizes in our
special steel company and its excellent logistics in the production facilities,
like rolling mill etc. with our logistics and processing centers all over the
world, where we will install 3D printing facilities for customized parts in
small lot sizes, providing the raw material for 3D printing. The service would
include post processing of CAD models, consulting to improve mechanical
properties, selecting the raw material quality and 3D-printing of the parts. But
we will also offer annealing and coating, eventually using nanotechnology
where appropriate. If a costumer needs many different machine parts in small
lot sizes we are ready and able to run a 3D printing facility at the costumer's
site. Thus we will manage to stay in market and to stay competitive, bundling
new services in the digitally disrupted market segments, while exploiting our

traditional skills in the left high volume markets like automotive industry etc."

Martha remembered how a few years ago they managed to do something similar with their equipment manufacturer for the maintenance of the rolling mill—only the other way round.

It would be another stage on their journey through the digital transformation.

Automation, fast cooperation with costumers, interaction with partners, help lines or web applications for consulting costumers—in a couple of years it will be hard to distinguish if we are talking or e-mailing with a computer—an Artificial Intelligence—or a human. Where are we heading for? How will we be taking decisions in future?—Who will be taking the decisions?

6.2 Hybrid Intelligences—Man and Machine Collaborating in Organizations

When we look in biology we see that a hybrid, also known as cross breed, is the result of mixing, through sexual reproduction, two animals or plants of different breeds, varieties, species or genera (Biology Online dictionary 2017). But we find the term "hybrid" also in mythology. Mythological hybrids are legendary creatures combining body parts of more than one species, one of which is often human. (this somehow reminds of Cyborgs in Science fiction but also of technologically enhanced humans with implants or prosthetic devices). In technology we find the term "hybrid" for example with vehicles using both combustion engines and electric power sources—Hybrid electric vehicles. The term hybrid is also applied in other domains like power generation etc.

So if we want to name the merging, collaboration and close coexistence of man and machine in the upcoming age of Artificial Intelligence and its collaboration with humans—thus bringing their intelligence into the organizations—it is appropriate to speak about hybrid intelligencies. So a hybrid intelligence can be defined as collective intelligence of humans and elements of Artificial Intelligence closely collaborating in order to serve the purpose of an organizational unit or an organization as a whole.

Let us assume that the primary design of such a hybrid intelligence is human work. The further development and evolution of such a construct may well be a process of emergence and self-organization and thus may be not merely human at all.

That is another reason why it is worth thinking about the subconscious mind of an organization—about an organization as a hybrid intelligence consisting of intelligent people and Artificial Intelligence, deep learning etc. up to artificially created AI by AI—hopefully under control of humans.

Let us discuss AI as an element of hybrid intelligences first: The poet and lyricist Setz (2014) has written an article in the feuilleton section of Die ZEIT and asked the question: Do voice-enabled computers need people at all? His answer is no—because androids can think for themselves. They are independent beings who have their own language and their own culture. Setz believes the future of Artificial Intelligence lies in the emancipation from humanity. "Language robots shall form a real foreign intelligence by themselves—regardless of the absurd demands of the human communication world".

If we with our human intelligence gain access to this (future) non-biological intelligence and "extend" ourselves, further exponential developments will occur—according to what Ray Kurzweil postulated for the information technology area, where he derived the LOAR ("Law of accelerated returns") from technological developments in the past decades.

With accelerating progress in brain research and computer technology (neuro-morphic computing and neural networks as well as brain-computer interfaces) the brain is on the way of being emulated with "reverse engineering". Thus the brain is about to gain access to its "source code" and will have the opportunity to further accelerate in iterative cycles to improve. With this expansion of our intelligence to Artificial Intelligence we will be able to achieve ever higher levels of abstraction. An ultra intelligent machine could still produce better machines—there is an explosion of intelligence to come. One of the main works of Ray Kurzweil "The singularity is near" (Kurzweil 2005) describes these scenarios. This interlocking of the real and virtual world is shown in the illustration in Fig. 6.1. ingeniously.

Where does this journey lead us to? And how fast does it go? What will the development of the human mind be? Who has access to these resources? Who influences the development? What about our consciousness? What about the free will of the individual, will it still be there? What is the identity of the individual?

These questions are not directly subject of this book. However, it is subject to how we position ourselves in our organizations and organize so, that we—organizationally and ethically—can handle these technologies (that have the potential for Artificial Intelligence) with our full awareness and in a way that we can consciously decide, what we let happen in the background—in the subconscious mind of the organization. So a lot of "Artificial Intelligence" may "creep in" the subconscious mind of our organizations. We are challenged to recognize that we have to design the hybrid intelligencies in our organizations actively. For the sake of the appropriate design of the subconscious mind of the respective organization we have to handle that properly.

Let us discuss the two poles of the thesis—an Artificial Intelligence, emancipating from humanity, as Clemens Setz postulated it on the one hand and our traditional image of an organization as a purposeful, more or less formalized cooperation of humans on the other hand: The intelligent algorithms that are capable of learning by themselves threaten to control us for example considering the

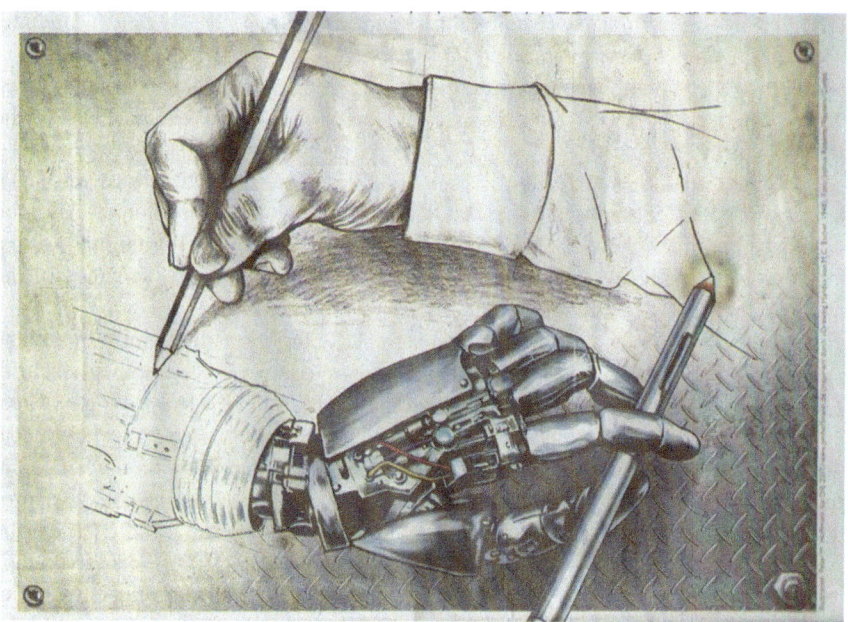

Fig. 6.1 The future?—hybrid intelligence? Die ZEIT, 13.2.2014

digitized home automation or the connected car—things some people obviously appreciate.

Yvonne Hofstetter says in her book "They know everything" (Hofstetter 2014) that there are three different types of Artificial Intelligence:

- Expert systems: expert systems are good to translate data into knowledge, I can, for example, find an error in the electronics of a car without the programmer explicitly anticipating this in their analysis program. The system recognizes the error itself.
- Swarm Intelligence: a population of autonomous software programs together solving a problem in cooperation.
- Optimizer: learning systems that continually improve autonomously, without human intervention.

The optimization of our daily lives, that intelligent machines promise us, requires constant monitoring of our lives. Apple Watch, Google Glass and other products and systems of augmented reality and expanded perception are the precursors of intelligent tools embedded in our life. The human will be embedded in the sensors, microphones, etc., which can not be switched off. We risk becoming dependent on them. Humans are increasingly losing control over how they are being assessed, classified and provided with information,

Let us imagine a scenario: *If, for example, the digitized home automation system based on the status messages in my phone recognizes that I am on the way home*

and starts to rev up the room temperature and humidity in accordance, then this
works only when I accept "uninterruptible total surveillance".

On the other hand we have traditional organizations relying completely on human ressources. Isn't AI a threat to the autonomy of the individual? Is the comfort worth it?

Intelligent machines are going to predetermine our future so that humans are in danger of losing the need and thus—longterm—their ability to make decisions. On the other hand intelligent machines will support us in our work. For the people, the question arises: Who will "have the last word"? Will we be able to "turn off the machine" in case of a machine-induced crisis? We need the organizations to keep going somehow—even without the machines. The issue of resilience (see Sect. 5.3.7) will gain importance in this context.

This raises the question: When is a machine actually intelligent?

One possible—pragmatic and easily understandable—response based on Hofstetter (2014) is: A machine or an algorithm is smart, when they show behaviors that were not provided by the programmer in this way. The system makes decisions that you have not thought through and defined in all its branches and consequences. In particular, systems that can develop a plan and a strategy probably have to be described as intelligent.

It is striking that up to now the application of Artificial Intelligence, for example, in self-driving cars, prostheses, robots etc. is often implemented "decentralized". AI is not all controlled—as comparable in humans—by the central "brain of the android". Even in systems with capabilities and features of Artificial Intelligence, the concept of distributed intelligence will be probably prevailing for some time. This distributed structure supports the analogy with "organizations as hybrid intelligence". Though—on the other hand—artificial intelligencies like Watson, depending on a huge amount of stored knowledge require more central architectures.

(Hopefully) assuming that the individual human will play the decisive role when man and machine are collaborating in our organizations it is worth to have a look how the usage of the new technologies by human individuals—outside their professional context as well as within the organization—is developing.

6.2.1 Man—The Human Individual and His Informational Patterns of Behavior

Facebook argues when asked about the impact of giants like Facebook on journalism that "it is our responsibility to provide our users access to a maximum of information and knowledge, which enables them to decide themselves" (according to Andrew Anker, at facebook responsible for journalism) (Müller-Wirth and Wefing 2016). Journalists and publishers can choose their topics and headlines and they can express their opinions in comments. They filtered, prepared and presented the information for their costumers. The costumers could choose their newspapers and its way to select and present information. (Remark: It was a renowned tradition of angloamerican journalism to clearly separate facts and opinion—an eroding concept completely impossible to live in social media due to its inherent mechanisms and structure).

Now Facebook and Google are deeply involved in the distribution of news and information—and they have an incredible coverage as their service is for free. So the journalists of the preferred newspaper are not the main information providers any more as Facebook and Google have massively changed the way people use online media and consume information. Now there are producers of news that were exclusively designed and programmed for this biotope. Like The *Angry patriot*, *American News*, *Being Liberal* etc. (Müller-Wirth and Wefing 2016), new "information providers" being more producers of opinions and less provider of information and news. They are completely meshing up opinion and facts with pictures, slogans, short comments etc.—far away from the tradition of good journalism. They definitely do not contribute to moderation and de-escalation of their readers and of society. Additionally AI- entities like chatbots can create multiple comments generated with AI, that can hardly be detected as originating from an algorithm and not from an individual human. This can massively influence the opinion building process in the social media and influence elections, propagate defamations etc.—the echoe chambers get literally filled with "garbage" generated by algorithms and not by humans who today easily can post "hate-postings" without being effectively persued.

That is the point when we will have to discuss whether to abandon anonymity from the internet of the future though anonymity is important in many situations. But there are possibilities to secure the important ability to give anonymous information and indicate crime or corruption for example via whistle-blower hotlines with advocates who guarantee anonymity for the respective person towards their employer or public authorities.

We increasingly experience the disintegration of information and messages into snippets and flashes—Facebook, Twitter and so on being the glorious examples and drivers.

Journalists and Scientists speak about hyper-personalisation, concerning

- How information is distributed (pushed to the consumer by algorithms who know exactly when you are regularly commuting in the train etc. being ready to receive information)
- How information is selected (algorithms selecting what is pushed to the consumer—with less informed consumers building their own "echo chamber" while using the "like-button")
- Some predict that in future the same story will be told personalized by the algorithms according to the interests, preferences and previous knowledge of the respective individual.

Who will control those algorithms following which patterns?—Who will pay for what? This will probably end up in massive manipulation and disinformation—finally endangering democracy.

I mention this at the beginning of this chapter about man+machine, about emerging or designed "hybrid intelligences" entering and servicing organizations and also forming the subconscious mind of our organizations, because it demonstrates the potentials for the bad and for the good of the respective organization. It might be used to manipulate your staff as well as to create benefit.

I am confident that this hyper-personalisation with the necessary algorithms will not be used within organization due to several reasons:

- Employees tend to have a different approach when acting privately and feeling free to choose compared with their approach in their professional organizations.
- Employers are subject to restrictions how to use personal data of their employees for example how these use the systems. The employers are not allowed to regularly extract behavioral patterns etc. at least in Western and Central Europe, Canada and Australia. There are less restrictions in the US and very little restrictions in other areas of the world.
- The algorithms and big data being the base for this hyper-personalisation are not open source and are only available for the giants in this business like Facebook, Google, Amazon etc.—at least for the time being and the near future.

So many employees will—though doing their job in their organization—maybe be captured in their private "echo-chamber", maybe they will be lacking education about reasonable use of information resources and social media and maybe they will be lacking the knowledge and the skills how to deal with that.

But there is a positive scenario concerning the rising use of AI-technologies within organizations with many employees being confronted with information retrieval and the use of AI for decision support in their job. They will have to be trained in the use of these technologies, they will more and more become aware of the patterns and algorithms behind, they will have to be trained, when to mistrust decision proposals on simulators (see Sect. 5.2.6) etc. Maybe that will be changing their attitude towards private information use as well and thus turn the employees more critical concerning forwarded news and information and making them aware of their personal echo chamber they are already in or on the way to. Maybe some are going to leave their echo-chamber and change their pattern of information use in their private sphere. So organizations can influence society positively when actively incorporating the new technologies with good information, training and participation of their employees. Thus the quality of the education system of our youth concerning the use of information will be the central playing field for winning the future and for stabilizing democracy and society.

So what should we consider when creating and forming hybrid intelligences in our organizations in order to facilitate collaboration between man and machines without losing control as humans and—unintentionally rendering the primacy to the machine? How to team up humans and AI for decision making and working?

6.2.2 Teaming of Human and AI

Let us get to the core of everyday work as well as strategic work in current organizations and look at what hybrid intelligencies might be able to perform and let us look at how AI elements and humans will behave in decision making when AI and humans are teaming up as hybrid intelligences.

Let us look at "narrow AI" first. Programs like *BlindType* compensate for human input error, and next generation phone-answering services convert your requests into commands. By assigning values to different situations, narrow AIs can make choices that maximize their rewards, an approach that let for example a robot figure out the best way through an obstacle course.

Artificial Intelligence is also getting better at analyzing large sets of data and synthesizing new data that fit the set, which we've seen in programs that write music or create new art. The path from "narrow AI" to "general AI" is likely to be a winding one. It is unclear how long it will take to create or raise learning machines capable of passing the Turing Test and which interact with us on a social level. There's little doubt that we will eventually have computers that equal and even exceed the human brain—at least with certain tasks ("narrow" AI) first and possibly "general" AI later. And they'll be doing a lot more than just taking our calls....

Starting with welding robots in cages segregated from humans to narrow AI we are moving on to true "artificial general intelligence". Now we can imagine that the collaboration of humans with AI in hybrid intelligence is on the way to change. Robots accompany doctors and nurses in hospitals, drivers are collaborating with AIs in self-driving cars and trucks. Robots, AIs and humans are teaming up. Will humans stay in the driver's seat at this stage of evolution?

So let us consider in the next sections various aspects and different views on this issue of how man and machine will collaborate in future and how hybrid intelligences and Artificial Intelligences will be embedded in organizations of the future.

One of the key questions from the human side will be—in a longterm perspective—whether there will be a role and work left for humans? Let us look at this first.

6.2.3 Will There Be Jobs Left in Self-tuning Organizations?

I don't want to answer this question from a mainly macroeconomical point of view but more from what we can expect on the level of each organization because that is what is going to be perceived as threat or benefit from the people acting in an organization—from people being or becoming part of an hybrid intelligence.

Jurdak (2016) offers the optimistic view that massive dislocations of workers are not inevitable. It will be key for the people who want to keep working, and the employers who have work to offer, to reframe the relationship of computers to humans as one of augmentation, not automation. Automation means starting with what the worker does today, and then subtracting—allowing machines to take over whatever tasks they can perform more cost-effectively. Augmentation starts with what the worker accomplishes today, then adding—spotting the possibilities for combining machine and human strengths to do what neither could do separately. But from my point of view one question remains: How to motivate profit-oriented organizations to invest in augmentation, when automation is more profitable? So augmentation (which can largely be associated with hybrid intelligence) will only win when the effort to be invested is lower than resulting productivity gains or/and when automation is already exploited from the profitability point of view and augmentation seems attractive to invest in, or augmentation produces new products and services with demand in the market, that cannot be accomplished with automation alone.

Brynjolfsson and McAfee (2014) have a more pessimistic view and say that the Second Machine Age is playing out very differently than the first Machine age did. The main technologies of the industrial revolution were, despite predictions to the contrary, complements to human labor: they greatly increased demand for the skills and capabilities of the average worker. The evidence so far, however, is pointing to a different conclusion about the digital technologies of the Second Machine Age. In aggregate, they're acting much more like substitutes for human labor, reducing demand for at least some types of workers' skills and abilities. So the human part of the hybrid intelligence will see the fast developing AI part as a threat. This will not be so relevant for organizations in competitive markets—they will have to improve their short term productivity with AI-parts to stay in market and to survive. But in a longterm view the way how to handle this feeling of being threatened by AI whilst taking the right strategic decisions for the organization will be a question of sustained economic survival—in case the organization fails to leverage the potentials of hybrid intelligence.

Something almost as interesting is emerging in another perspective of decision making: How are organizations automating decisions? Is there a way for organizations to make frequent, calibrated adjustments to their business models, their resource allocation processes, and their organizational structure—and to do this without direction from the top? The organizations might probably act very decentralized and they might base their decision making on a close observation of two themes—depicted by the following examples (following Boston Consulting Group—Reeves et al. 2015).

- First—automated adjustment of product offerings: Companies like Netflix and Amazon are now extraordinarily good at adjusting their product offerings automatically, in real time, in response to individual customer behaviors. It's true that they use algorithms to do that work. But the processes and technologies underlying those algorithms aren't magic: it's possible to pull them apart, learn what makes them tick, and use that learning in other settings.
- Second—transfer the principles and algorithms to other business areas: Some of the same companies have started to do just that: apply the principles underpinning these algorithms to managing other areas of the enterprise. Extending

beyond bits and bytes, these principles offer fresh insights on how to manage enterprises in dynamic and unpredictable environments, as Boston Consulting Group (BCG) reports—speaking about the "self-tuning enterprise" (Reeves et al. 2015).

BCG looked at how organizations are applying what they've learned about self-tuning—in the form of continuous, self-directed experimentation at the enterprise level, using Alibaba as a case study (Reeves et al. 2015).

What will "self-tuning enterprises" as wide and abstract form of hybrid intelligences mean for management? What will be left to manage?

6.2.4   What Will Be the Impact on Management Structures?

According to Boston Consulting Group (Reeves et al. 2015), the basic idea is: *don't try to manage what is better left to market mechanisms*. As a young enterprise in a fast-changing environment, Alibaba had been trying not only to institutionalize change but also to bring the marketplace into the organization. Adapting to change through managed experimentation had already become a popular idea and a reality for many "net-native" companies. But Alibaba was taking it a step further by trying to continuously update not only its product offering but also the elements of the business that we might ordinarily assume are fixed: the vision, business model, organization, and information systems.

BCG had, in parallel, been modeling the effectiveness of different approaches to strategy and execution in different environments. They carried out this modeling using a so-called multiarmed-bandit algorithm—coincidentally, the same sort of algorithm that Alibaba and other digital-market players use to recommend products to customers. These algorithms turn out to be extraordinarily good at rapidly tracking and stretching customers' changing needs—tuning themselves into changing circumstances. They do so—following the BCG-article—by operating three learning loops in a rapid, automated fashion: *adapting* to changing customer needs, *modulating* the exploration of new needs, and *shaping* the emergence of new needs.

So BCG decided to marry the two streams of thought and codify what Alibaba had been implicitly trying to do for some time: build an enterprise that could tune itself at all levels to changing circumstances. BCG tried to derive a set of principles and actions that any organization could use to create a self-tuning enterprise. The basic idea was to replace elements of a company's business system—elements that are usually fixed or specified in a top-down fashion—with self-directed mechanisms that continuously evolved, guided by and shaping the marketplace. That is a good example of how to deal with hybrid intelligences—how to shape them, being aware of what will run in the subconscious mind of an organization and how the conscious mind interacts with it and how it is influenced or even controlled by the subconscious mind.

Algorithmic competition seems to become an increasingly important aspect of business. In this case, we are not focusing on machines replacing or supplementing man. Rather, we are highlighting the importance of algorithmic thinking for updating some key business concepts like vision, organization, business model, and information systems that were framed in simpler, more stable times. The metaphor of the subconscious mind of the organization also offers a concept of learning and not only of actively shaping, forming and enabling an organization.

Who needs self-tuning? Any enterprise that, like Alibaba, that faces a complex and dynamic environment requiring readjustment at all levels. And that's an increasing number of not only "net-native" companies but also of traditional organizations that may be facing disruption from new technologies or competitors.

As we see in this example, the conceptual view of an organization as hybrid intelligence might evolve—concerning the organization's decision making—already as far reaching when creating a self-tuning enterprise with powerful algorithms. The processes and algorithms of these self-tuning organizations are supposed to react on what is happening in the market. But this hopefully is performed by a hybrid intelligence with human control in the sense of accepting, rejecting, replacing or modifying elements of the organization's business system. That is where the appropriate shaping of the subconscious mind of an organization as management task becomes important. Referring to the chapter before—shaping the subconscious mind of an organizations—it seems to be appropriate to form flat organizational structures, "edge-based" organizations (see Sect. 5.3.4), oriented towards the markets and agile and easy to adapt in order to come near "supervised" self-tuning organizations by managed experimentation. We need "fluid" organizations and organizational units in order to enable flow and the growing of ideas and provide enough cohesion and stability to make that growing happen-in spite of the dynamics and the ever accelerating economy evaporing good novel ideas and concepts too early.

6.2.5 Real Life (RL) Gaming Tears Down the Walls and Is Bringing the Stories Back

How can this idea of managed experimentation be supported? Now let us view the technologies that are "tearing down the Walls between the Digital and the Real", as Jason Ganz reported (Ganz 2016). Gaming gives us an impression of what is near in our hybrid intelligence settings in our organization: Integrating augmented reality (AR) as a primary platform to perceive and analyze what is happening, we will fundamentally dissolve the walls dividing the physical and the digital. We will have to adopt the best learnings from Artificial Intelligence to create scenarios which are surprising, delightful and deep—combining real life locations with fully immersive virtual reality experiences. Look what Gaming VR (Virtual Reality) and AR are offering combined with AI. Real Life (RL) Gaming is possible.

Ganz (2016) writes that the most important thing he remembers about the stories that he was told when he was young was, that they were deep. "Even more, they were wild. We didn't know all of the rules. And to be honest, we didn't really want to".

"I am not sure whether such deep stories are possible with games today. Our games are wide and expansive, with massive worlds to explore. But they're flat—fairly predictable systems running off of predictable human algorithms. RL games are the same way, running off of the same predictable experience". Jason Ganz predicts a different future of Real life gaming—leading to the "deep stories" that impressed him in his youth (Ganz 2016):

- **Augmented Reality Makes It Real**: No one is going to be playing Pokemon Go or any other RL game on their phones once AR headsets hit the market. By dropping the smartphone and integrating augmented reality as the primary platform, we will fundamentally dissolve the walls dividing the physical and the digital.
- **Machine Learning Powered Magic**: RL games will adopt the best learnings of Artificial Intelligence to create scenarios which are surprising, delightful and deep. We'll really feel like we are exploring wild, vibrant worlds rather than the wide but flat experiences of today.
- **Customized VR Locations**: Augmented reality headsets will be great for the day-to-day operations of RL games. But sometimes you're going to need to go on special quests and missions which require more immersiveness. Luckily, VR theme parks like The Void will fill this gap, combining real life locations with fully immersive virtual reality experiences. You'll wear top-of-the-line VR gear and drop into hand-crafted virtual experiences that bring the game fully to life.
- **Deep Social Ties**: Many people will play these RL games on nights and weekends with their friends. We'll roam around our cities, adventuring and socializing. The bond of the game will make it much easier and more common to interact with strangers, and there will be strong communities within each reality (although, people being people, this will also lead to some malicious actions).
- **Physical and Mental Benefits:** The physical benefits are obvious—these games will require you to be at least a little active. The mental benefits will be more subtle, but probably more important in the long run. Gaming is already being tested as a treatment for depression and anxiety. RL Games will be extremely effective for this. In fact, there are already several posts on Reddit from people who say that Pokémon Go is helping them battle depression and even suicidal thoughts.

Jason Ganz reports that he personally had incredible, authentic, and most importantly, human interactions thanks to Pokemon Go—quite similar to reports in newspapers and magazines.

From my point of view there are many threats to be considered with Real life Gaming for the humans acting: I am traceable, the games immerse in my life. What is the monetizing model behind?—Is it only location based advertising? Is it conditioning my subconscious mind?—Is it subtly conditioning the subconscious mind of my organization?

In fact with things like AI and machine learning, the experience can become a lot more life like. This is already seen for example in some football games where the computer learns from the way you play and changes to counter this. But clearly there is much more potential. Games are often too linear and they don't often go the way you'd like them to. The time will come when a game can observe your actions and reactions and adapt the story line to fit.

But there are two contradictory points to be mentioned. Firstly, gaming can be addictive and can be used by people who wish to withdraw from social interaction e.g. 2nd Life. Secondly, is virtual reality a fad? 3D cinema doesn't seem to have changed cinema in the way some said it should, as the story and characters are more important.

"With all the technical development in games, what is the most popular game? It's something really simple like "candy crush" or remakes of "Super-Mario"", Jason Ganz says. Obviously people like it because they can use it to switch off for 5 min. This is the opposite of a deep immersive game where you need hours and to be in the right frame of mind.

Envisioning AR-lenses in the eye, brain implants, EEG headsets as Brain Computer interface to control the cursor or move objects or-in future-to read your mind and predict what you are heading for, we can imagine a virtual layer between our eyes and the real world, that is going to interpret for us what is really going on in real life. We might be "outsourcing" and losing our sovereignty for interpretation.

But that's where we're probably heading for—with digital natives taking control in our organizations. But RL gaming isn't just the future. It's here today. What will be the impact concerning hybrid intelligencies in our future organizations? It is possible to transport procedural skills and improve procedural memories and to transport "stories" with tools (AR, VR, Gaming) that allow deep immersion, which support that. It is bringing "the stories back" in terms of what impressed and framed us in earlier years (see the view of Jason Ganz as described above). Let us consider to use these RL technologies for simulations, trainings, acquiring knowledge, augmented video conferences etc. thus providing conviviality and maybe making it easier to give empathy across space. It will probably change the intuitive capability of individuals too. In any case these technologies—consciously used—can provide bonding between the human and AI parts of an hybrid intelligence. It will also provide opportunities to help shape and form the subconscious mind of an organization. There is some potential to accelerate and improve decision making in such an environment of a hybrid intelligence.

6.2.6 Blending Analytics with Intuition

Let us look at one basis of decision making—the data available. Let us also look how to process them, how to present them, so that good decisions can be taken including the experience and skills of those responsible for the decisions to be taken? MIT has analysed analytic work profoundly (Ransbotham 2016) and worked out four key issues:

- **Data awareness and responsibility**. As managers rely more on data, their awareness of data in the organization where it is, who has it, what's available, how to find what one needs—has to grow as well.
- **Openness to new ideas**. Entertaining a wide range of ideas is fundamental to cultivating both innovation and competitive advantage with analytics, but creating room in an organization to enable that to happen demands openness to new ideas that challenge the status quo, along with a tolerance for mistakes. "Analytical Innovators use existing data to create or curate new data by looking at it in inventive ways, developing new attributes, or asking questions in new ways".
- **Signals about the importance of analytics**. Employees frequently look for signals about what is important to management, and whether what is important today will be important tomorrow. Establishing organizational structures such as data councils, data labs, and centers of excellence signals to staff that the organization is taking data seriously as a core asset. Senior managers who use analytics themselves and set clear expectations about staff's use of data in proposals make visible statements about the importance of analytics.
- **Decisions that blend analytics with intuition**. Managers in more advanced analytics companies give more weight to analytics when they make key business decisions. Equip senior managers with skills and the attitude to appreciate that analytics can take intuition much further, in some instances, than intuition by itself. Help managers appreciate that the process of developing analytics is not a mechanical process devoid of intuitive leaps. Blending analytics with intuition in decision making can produce more effective results than either alone, especially when making strategic decisions.

To blend analytics with intuition—shaped and provided by the subconscious mind of the individual—will be a key feature of hybrid intelligences and gives humans the ability to control the outcome. Intuition based on experience and envisioning the future—maybe supported by managed experimentation—will be a key ability to cope with increasing complexity and uncertainty. Intuition will be one key feature that will help us as humans to "stay in the drivers seat"—along with empathy, conviviality and our human capabilities in being creative and having the right "feeling". But—for making good decisions and acting wisely—we will have to be able to blend these capabilities with what AI will provide us, thus forming human-controlled hybrid intelligencies. So how shall we organize best around our core activity in an organization—decision making?

6.2.7   How to Reorganize Around Decision Making—What Will Be the AI Part?

A functional organization is only partly determined by structure. Equally important is the web of relationships that people develop over time to get things done.

Reorganizations disrupt those relationships, hindering productivity until connections can be rebuilt within the new structure. As one senior executive of a multinational IT organization told me once: "Every time we reorganized, we lost at least a year of innovation and invested high efforts to explain it to our costumers, partners and employees."

Following Ashkenas (2011) you should ask 2 questions before starting to reorganize:

- What's the problem you're trying to solve?

If the structure has become overly complex such that accountability is diffused, etc. you should ask:

- Is reorganization the only possible solution?

In reality, a company's structure results in better performance only if it improves the organization's ability to make and execute key decisions better and faster than competitors. If you can synchronize your organization's structure with the decisions that have to be taken in daily business as well as strategically, then the structure will work better and performance will improve.

To reorganize around decisions, Ashkenas suggests to focus on six steps.

- Identify your organization's key decisions.
- Determine where in the organization those decisions should happen.
- Organize the macrostructure around sources of value.
- Figure out what level of authority decision makers need.
- Align other elements of the organizational system, such as incentives, information flow, and processes, with those related to decision making.
- Help managers develop the skills and behaviors necessary to make and execute decisions quickly and well.

Chrysler, for instance, reorganized its operations three times in the three years preceding its bankruptcy and the collaboration with Fiat. Each time, executives proclaimed that the company was on a new path to profitability. Each time, performance didn't improve (Ashkenas 2011).

An organization's structure, similarly, will produce better performance if and only if it improves the organization's ability to make and execute key decisions better and faster than competitors. It may be that the strategic priority for your organization is to become more innovative. In that case, the reorganization challenge is to structure the company so that its leaders can make decisions that produce more and better innovation over time.

For most companies, this requires a fundamental rethinking of their approach to reorganization.

Instead of beginning with an analysis of strengths, weaknesses, opportunities and threats, structural changes should start with what Ashkenas calls a "decision audit". The goals of the audit are to understand the set of decisions that are critical to the success of your company's strategy and to determine the organizational level at which those decisions should be made and executed to create the most value—

answering the six questions above. As you conduct your own decision audit, you need to consider two types of critical decisions:

- Big, one-off decisions that individually have a significant impact
- Small, routine decisions that cumulatively have a significant impact.

If you can align your organization's structure with its decisions, then the structure will work better, and your organization's performance will improve. Ultimately, an organization's value is no more (and no less) than the sum of the decisions it makes and executes. (This resembles the pattern proposed for knowledge management with decision and action in focus—see Sect. 3.2.12 and Fig. 3.1). Its assets, capabilities, and structure are useless unless executives and managers throughout the organization make the essential decisions and get those decisions right more often than not.

Try to assess—in this decision audit—your decision quality (whether decisions proved to be right more often than not), speed (whether decisions were made faster or slower than competitors), yield (how well decisions were translated into action), and effort (the time, trouble, and expense required for each key decision).

According to Ashkenas research revealed no strong statistical relationship between structure and performance. Survey respondents' views about the structure of their company were not an accurate predictor of either decision effectiveness or financial results. So, when reorganizing a company, decisions rather than structure should be the focus. Ashkenas strengthens these hypotheses with cases from Ford and Amazon.

If you find that you are already focused on the right set of critical decisions and you find that they are placed where they should be within the organization, then the source of performance problems is unlikely to be structural. Problems may stem instead from other organizational issues. Perhaps there is an incentive problem with the sales force. Maybe leaders place too much value on building consensus, at the expense of decisive action. Maybe you lack decision support from your information systems. Whatever the issues, they can probably be fixed without a wholesale structural redesign and all the effort and trouble around.

What will be **the AI-part in organizing hybrid intelligences around decision making**? Will it be restricted to decision support? Will it be automated decision making as described regarding the shaping and forming of the subconscious mind of organizations in the above chapters of this book?

According to Fan (2016) citing different sources in "Will Letting AI Make Our Decisions Be the Best Decision We Make?", "the next big breakthrough in design and technology will be the creation of products, services, and experiences that eliminate the needless choices from our lives and make ones on our behalf, freeing us up for the ones we really care about: Anticipatory design."

In a nutshell, anticipatory design creates an ecosystem where users don't have to make decisions—instead, a choice is made automatically on the user's behalf without input. Here, an AI determines the best option based on the user's prior behaviors, preferences, and other data, guided by some simple business logic and common sense to smoothly complete the decision cycle. In anticipatory design for example, an AI assistant scans for upcoming out-of-town events in your calendar and if you decide for one automatically books a ticket, carefully choosing airlines, seats, flight time and price based on previous bookings. I am convinced this is going to go deeper and in more substantial features of organizations than this easy-to-understand example. It deeply affects sovereignty and autonomy to push more and more in the realm of the subconscious mind of organizations.

However, like any other machine learning system, the more you use it, the better it gets. In the end, the system simplifies life by removing intermediate steps toward a goal.

According to S. Fan's article, some pioneers are already taking baby steps in that direction, though with mixed results. For example, Amazon, Netflix, and Pandora offer a smattering of recommendations based on a user's past picks—though one might argue those systems often make things harder because they still require the user to make the ultimate decision.

In contrast, a Nest smart thermostat does all the work without consulting you. This "Internet of Things poster child" automatically tweaks room temperature based on the time of day and your prior preferences.

How should we build a Decision-Driven Structure?

All complex organizations must be broken down into manageable pieces to ensure that roles and responsibilities for making and executing critical decisions are clear. To get the holistic view identify AI's role in decision making as described above.

According to Blenko et al. (2010) it is wise to compensate the disadvantages in each established structure or chosen structure for reorganization: The leadership team might create corporate councils and boards directly under the company's operating committee for example to prevent to loose costumer focus in case you decided for a structural design around corporate functions. These cross-functional groups formulate and evaluate alternatives for each of the organization's major strategic initiatives and then make recommendations to senior management. This process can accelerate decision making without sacrificing decision quality. The structural overlay of councils and boards can help functional leaders to collaborate and make effective decisions about budgets and resources.

Creating parallel decision-making authorities may appear to be in conflict with the general principles of simplicity and clear accountability, but this approach may lead to a more streamlined process. Such organization overlays like councils and boards can introduce valuable expertise that formal structures cannot easily accommodate. They allow fewer people to be involved in making and executing critical decisions—in effect, reducing the number of decision nodes.

The result which executives always seek from reorganizations is yet so seldom accomplished: improved performance. One reason is the complexity of what is to manage. Even after reducing complexity as far as possible there remains enough you have to be aware of in order to organize in the best possible way trying to drive your organization's performance.

6.2.8 *What Does Really Drive Business Performance?*

Being well and appropriately organized around decision making is one prerequisite to be successful concerning the purpose and the goals of the organization you are part of—maybe already being some sort of or including hybrid intelligence. What will drive our business performance in such settings?

For example, academics and consultants such as Robert S. Kaplan and David P. Norton, the developers of "the balanced scorecard," have encouraged managers to hypothesize causal links and to develop strategy maps to identify the key drivers of financial performance in their organizations. The problem is that developing strategy maps on the basis of managerial hypotheses means the maps are constrained by managers' prior views of what drives performance. Managers often make assumptions about the relationship between, for example, customer loyalty and profitability, even when the presumed links haven't been fully tested.

According to Silvestro (2016), "Topology mapping" is a well-established approach for depicting complex networks and is commonly used to represent the relationships in settings such as transportation and communications networks. This approach can be adapted to represent networks of performance linkages in order to help managers visualize and explore the sometimes complex relationships between the various aspects of business performance. It requires managers to identify the key performance indicators (KPIs) relevant to their business, measure the correlations between the various KPIs, and then build maps of the positive and negative correlations. Sylvestre describes **Seven Steps of Performance Topology Mapping**:

1. Identify the most important key performance indicators (KPIs) for your business.
2. Measure the correlations between all the pairs of KPIs, and identify the statistically significant positive and negative correlations.
3. Build a topological map where the KPIs are the nodes and correlations are represented by connectors. Use different colors to distinguish between positive and negative correlations.
4. Use bold connectors to highlight particularly strong correlations. This may reveal a pathway of linked measures.
5. Identify the nodes that have multiple connectors. These may be key drivers that can be leveraged to improve performance.
6. "Morph" the topology map into configurations that provide a fresh perspective on the drivers of performance. The aim is to make the map easy to interpret.
7. Confront and explain any unexpected or uncomfortable findings. Are there unexpected correlations—or any links missing that you expected to find? Ask yourself: What are the implications of the correlations you see for corporate strategy and for the design and delivery of your products, services, and processes?

One key prerequisite for decision making and for topology mapping on the KPI-level is analytics. Ransbotham et al. (2016) in a research report by MIT Sloan

Management Review remarks that the term "analytics" refers to the use of data and related business insights developed through applied analytical disciplines (for example, statistical, contextual, quantitative, predictive, cognitive, and other models) to drive fact-based planning, decisions, execution, management, measurement and learning.

Five key findings came from his research:

- Competitive advantage with analytics is waning. Increased market adoption of analytics levels the playing field and makes it more difficult for companies to keep their edge.
- Optimism about the potential of analytics remains strong, despite the decline in competitive advantage. In addition, use of analytics for innovation remains steady.
- Achieving competitive advantage with analytics requires resolve and a sustained commitment to changing the role of data in decision making. This commitment touches many aspects of organizational behavior, from revamping information management to adapting cultural norms.
- Companies that are successful with analytics are much more likely to have a strategic plan for analytics, and this plan is usually aligned with the organization's overall corporate strategy.
- Most companies are not prepared for the robust investment and cultural change that are required to achieve sustained success with analytics, including expanding the skill set of managers who use data, broadening the types of decisions influenced by data, and cultivating decision making that blends analytical insights with intuition.

These findings strongly support the idea to focus analytics on the right questions (for example derived from topology mapping), on what you can actually measure, on what seems feasible from newly available "big data" and on what you think is driving your performance in a volatile ever-changing environment. Those targeted analytics—enriched by and embedded in decision support and AI capabilities—have to be well woven in the fabric of the hybrid intelligencies emerging in your organization. So the managerial skills required are changing too. This is even aggreveted by cross-organizational automated business processes towards the costumer.

6.2.9 Cross-Organizational Business Processes Incorporating Artificial and Hybrid Intelligences

Elements of Artificial Intelligences often already are and in future increasingly will be linked to process chains spanning several independent organizations. It is this what meets costumer's needs, with the costumer merely interested in the service and the product and not at all interested in the organization or the organizations behind.

Let us consider an example: We already see business travel planning services which automatically plan a business travel from A to B at a given date—automatically composing taxi (for example Uber), flight, accommodation etc. to the preferences of the costumer—maybe already known from prior business travels—thus giving proposals to be decided by the costumer. Artificial and Hybrid Intelligences are collaborating cross-organizational and cross-functional probably accompanied by a virtual assistant—chatbot—as interface to the costumer.

Thus the "subconscious mind of an organization" has interfaces to other organizations and their "subconscious minds" that work following their sets of rules, maybe self-learning without or with only little human intervention—but hopefully well monitored by humans and controlled when necessary.

Thus hybrid intelligences collaborating interorganizational emerge with arising questions concerning responsibility, liability of involved organizations—responsible for their respective part of this interorganizational hybrid intelligence representing a cross-organizational business process towards the costumer. Another managerial challenge.

6.2.10   What Are the Managerial Skills Required?

Exploring how technology is reshaping the practice of management, Lynda Gratton, a Professor of Management Practice at London Business School makes three predictions about management and technology in the FrontiersBlog published by MIT Sloan Management Review (Gratton 2016):

- "First, it was clear to me that the role of management as a coordinator of work would come under increasing pressure". The constant march of robotics and machine learning and the "hollowing out of work" makes management a more and more unclear practice. What is a manager, and what is it that they do? Are we witnessing the end of management?
- "Next, I could see the inevitable shift from a parent-to-child way of looking at the relationship between the manager and their team being questioned and ultimately superseded by an adult-to-adult form. The nexus of this more adult relationship relates to how commitments are made and how information is shared". When technology enables many people to have more information about themselves and others, then it's easier to take a clear and more adult view of the world. Self-assessments, particularly those that enable people to diagnose what they do and how they do it, can help them pinpoint their own productivity issues. They have little need for the watching eyes of the manager.
- "Third, it seemed to me obvious that technology would tip the axis of power from the vertical to the horizontal. Why learn from a manager when peer-to-peer feedback and learning can create such stronger lateral paths of coaching?" Moreover, technology enabled social networking is capable of creating more

robust and realistic maps of influence and power—"so no more hiding behind fancy job titles."

What do these three areas of change and risk have in common? They are all fundamentally about management. This requires a very complex form of management—managing virtually rather than face to face; managing when the group is diverse rather than homogenous; managing when the crucial knowledge flows are across groups rather than within. These are highly skilled roles in terms of both managerial capabilities (for example, to build rapid trust, coach, empathize, and inspire) and management practices (for example, team formation, objective setting, conflict resolution). "It is these managerial skills and practices that will be augmented by technology over the coming years in ways we may have not yet grasped but which will emerge, just as the use of personal technology has emerged." Linda Gratton says.

So—facing that AI parts will (starting with decision support) be "players" in decision making processes—for the formation and development of hybrid intelligences several conclusions can be extracted for the human parts of organizations:

- A shift in the role of management from "parent-child" to an "adult-adult" relationship is required.
- The parenting role should focus more on the AI parts of the organization
- This leads to more self assessment than watching managerial eyes on the employees,
- The more peer-to-peer feedback and learning is enabled by social networking, this will create new maps of influence and power which have to include AI parts wisely.
- This requires managerial skills to manage knowledge flows across groups rather than within, also considering the new AI-entities.

The rise of Uber has everyone excited about platforms and how they can create a fertile place for new businesses to be built and act as a conduit for flexible working. To build, grow and manage platforms—not in a merely technological sense but following the idea of a hybrid intelligence—will be a key managerial challenge to succeed in the digital transformation. In its specific path through this jungle called digital transformation, this—above depicted—form of managerial attitude and skill paired with a clear vision of the Big Picture, the respective organization is moving to, will enable good governance of hybrid intelligences and help the humans in an organization help to stay in the driver's seat. Thus organizational behavior of the organization and the individuals in the organization is going to change. How can that be controlled?

6.2.11 Conditions Are Decisive in Organizational Behavior

Let us have a look at what we can derive from the science of organizational behavior for decision making in hybrid intelligencies. It is obvious that—in these regards—we have to focus more on the behavioral aspects of organizational units and organizations as a whole than on individual behavior of individual participants. That is why the article "From causes to conditions in group research" by J. Richard Hackman seems very relevant to me (Hackman 2012):

Hackman remarks that—whilst we see reasonably widespread acceptance of complex systems ideas such as emergence and self-organization, there has not been much progress in using the concepts and methods of complex systems theory in empirical research on groups so far. So he suggests another alternative to traditional causal models of group behavior and performance. His basic idea is to move thinking about group behavior from a focus on cause–effect relations to an analysis of the conditions under which groups chart their own courses. He refers to studies of civilian and military flight deck crews. For these teams for example on approach to landing it is decisive to have conditions in place such that the natural course of events leads to the desired outcome—in this case, a good landing. In Hackman's view the same way of thinking applies in a wide variety of other domains of human endeavor like economics and medicine.

In the following you find a summary of six enabling conditions, Hackman sees as decisive:

- Real team.
- Compelling purpose.
- Right people.
- Clear norms of conduct.
- Supportive organizational context, concerning

  - the reward system,
  - the information system
  - the organization's educational system.

- Team-focused coaching.

As new forms of groups continue to proliferate—in combination with AI forming hybrid intelligencies—it may be a good time to rethink how we conceptualize, analyze, and work with them and which conditions we provide so that they can prosper. This topic is also a building bloc in the shaping of the subconscious mind of organizations especially concerning the "soft infrastructure" (see sect. 3.1.1). Team conditions for flight deck crews—as Hackman observed them—seem to be a good example, as they are accustomed to collaborate and live with advanced IT Systems (coming near AI), like autopilot etc.

Another example that should not serve as role model but as a lesson what can go wrong are investment-bankers, the way they worked and which conditions were set until the crash of 2008. As Nils Ole Oermann described in his book mainly dealing with building up the investment banking branch of Deutsche Bank (Oermann 2013), those investment guys were driven by three motivations:

- enjoy being the winner in the competitive playground,
- receive high bonus payments and
- to feel being part of an elite.

The behavior of these organizations was excellent from the short-term corporate view but disastrous from the societal and longterm macroeconomic view. (It was also disastrous for the longterm development of Deutsche Bank as we witness in autumn 2016). The conditions set by the Management of Deutsche Bank were obviously wrong. Collaborating with advanced algorithms and exploiting inappropriate legislative rules and lack of corporate rules they ignored all ethic rules and completely lacked the idea of being a part of a community or even serving a community—although all of them were highly intelligent in their specific framing.

It is frightening to imagine such people establishing and exploiting advanced AI in a non-regulated and uncontrolled mode, though it is not unlikely to happen this way due to the fact that regulation usually is too slow.

So let us turn now from decision making to a higher level of collaboration and coexistence and look at further aspects of hybrid intelligencies and in particular the role of humans in this setting.

6.2.12   Conviviality and Empathy as Guiding Principles Creating Hybrid Intelligences

The austro-american anthropologist Ivan Illich has defined the term conviviality, regarding how humans coexist with technical progress. This concept—though from the 70s of the last century—may help us to explore hybrid intelligences a little deeper.

Conviviality

In Tools for Conviviality (Illich 1973) Illich generalized the themes that he had previously applied to the field of education in his book "Deschooling Society": the institutionalization of specialized knowledge, the dominant role of technocratic elites in industrial society, and the need to develop new instruments for the reconquest of practical knowledge by the average citizen. He wrote that "elite professional groups have come to exert a "radical monopoly" on such basic human activities as health, agriculture, home-building, and learning, leading to a "war on subsistence" that robbed peasant societies of their vital skills and know-how. The result of such economic development is very often not human flourishing but "modernized poverty, dependency, and an out-of-control system in which the humans become worn-down mechanical parts." Illich proposed that we should "invert the present deep structure of tools" in order to "give people tools that guarantee their right to work with independent efficiency."

As reported by Smith (2001) the book's vision of tools that would be developed and maintained by a community of users had a significant influence on the first developers of the personal computer, notably Lee Felsenstein, who was influenced in his philosophy by the works of Ivan Illich, particularly by *Tools for Conviviality*. This book advocated a "convivial" approach to design which allowed users of technologies to learn about the technology by encouraging exploration, tinkering, and modification.

Illich worked to open new possibilities. He argued that we need convivial tools as opposed to machines. "A tool may have many applications, some very different from its original intended use. A tool may be thought of as an expression of its user. The opposite of this is the machine, where humans become its servants, their role consisting only of running the machine for a single purpose".

> I choose the term "conviviality" to designate the opposite of industrial productivity. I intend it to mean autonomous and creative intercourse among persons, and the intercourse of persons with their environment; and this in contrast with the conditioned response of persons to the demands made upon them by others, and by a man-made environment. I consider conviviality to be individual freedom realized in personal interdependence and, as such, an intrinsic ethical value. I believe that, in any society, as conviviality is reduced below a certain level, no amount of industrial productivity can effectively satisfy the needs it creates among society's members (Illich 1973).

Regarding hybrid intelligencies we should think about if and how to integrate specific AI in this network. The ideas of Ivan Illich concerning the problematic technocratic approaches should be well considered. As we all speak and write about man and machine, it seems to be more appropriate and wise to talk about "man and his tools" as synonym for this convincing idea of conviviality in this context of hybrid intelligence. As already remarked in the preface of this book our cultural evolution is an evolution of our tools in a wider sense. Look at Fig. 6.1 and imagine the saying from Robert K. Logan "We shape our tools and then our tools shape us".

I would suggest that "real" conviviality in the sense of sociability should remain a privilege for human intelligence or the human part in hybrid intelligences and thus help humans to exchange on an informal level—allowing emotions like joy and anger—because at least it is all about benefit for humans and their sustainable future.

But we have to equip the humans in organizations to stay in the driver's seat and apply AI in a sensible way in their hybrid intelligencies—as tools. Again Ivan Illich with his ideas about "Deschooling" opens the door to knowledge management on a very human foundation. M.K. Smith, who reviewed Ivan Illich's work argues for learning webs (Smith 2001)

Learning webs—new formal educational institutions. In *Deschooling Society* Ivan Illich argued that a good education system should have three purposes:

- to provide all that want to learn with access to resources at any time in their lives;
- make it possible for all who want to share knowledge etc. to find those who want to learn it from them; and
- to create opportunities for those who want to present an issue to the public to make their arguments known.

Note that these ideas were created long before the internet we know today even was on the horizon.

Illich suggests that four distinct channels or learning exchanges could facilitate this. These he calls educational or learning webs. "Educational resources are usually labelled according to educators' curricular goals. I propose to do the contrary, to label four different approaches which enable the student to gain access to any educational resource which may help him to define and achieve his own goals":

- **Reference services to educational objects**—which facilitate access to things or processes used for formal learning.
- **Skill exchanges**—which permit persons to list their skills, the conditions under which they are willing to serve as models for others who want to learn these skills, and the addresses at which they can be reached.
- **Peer-matching**—a communications network which permits persons to describe the learning activity in which they wish to engage, in the hope of finding a partner for the inquiry.
- **Reference services to educators-at-large**—who can be listed in a directory giving the addresses and self-descriptions of professionals, paraprofessionals and freelancers, along with conditions of access to their services. Such educators could be chosen by polling or consulting their former clients.

That's what knowledge management in modern organizations should be all about and that's what would be—from the technological side—already available with social media, search engines etc. If humans want to stay in the driver's seat of hybrid intelligences, organizations will have to provide this with a good contextual access to help its members to find what they need. On the other hand the organizational culture must be developed towards sharing of knowledge in order to exchange and get access to skills of the hybrid intelligence—notably its human elements. Microlearning as described above (Sects. 5.2.8 and 3.2.11) seems to be a very promising approach in this context.

Empathy

Let us now have a look at a basic human capability that we have to consider when speaking about hybrid intelligences—empathy. Jeremy Rifkin says in his book "the empathic civilisation" (Rifkin 2009) that humans are not soft-wired for aggression, violence, self-interest and utilitarianism but humans are basically softwired for sociability, attachment, attention and companionship. That basically sounds good for creating of AI and integrating it in the civilizations with organizations being pioneers in creating hybrid intelligencies.

For Rifkin this hypothesis can be derived from brain-MRI's for example, when you see a spider creeping up your companions arm, the same patterns show up in the MRI as if the spider is creeping up your own arm. Biologically the "mirror neurons" seem to be the reason for it. Sociologically it is the drive to actually belong—an empathic drive, Rifkin says. We show solidarity with our compassion. "Empathy is grounded in the acknowledgement of death and the celebration of life", Rifkin says. "Empathy is based on our frailties and imperfections".

So when we talk of building an empathic civilisation we talk about solidarity. Rifkin asks the questions "can we extend our empathy to the entire human race? Can we extend it to our biosphere and our planet?" Empathy is the invisible hand that drives us. In early times communication only extended to the local tribe, the creature from behind the next mountain already being an alien. Empathy only extended to blood ties. When mankind entered the hydraulic and agricultural civilization it allowed us to extend the "central nervous system" of mankind. Mankind started to de-tribalize and began associations with religious ties. With industrial revolution this extended to the nation state. For example the French were becoming the "extended family" of all frenchmen. The nation state is clearly fictious. "Now loyalties based on a complex energy and communication revolution annihilate time and space", Rifkin says. Following this development from tribal ties to religious ties, to nation states we might ask: "Can we connect our empathy to a single race in a single biosphere?" "If it's impossible to imagine that, then I can't see how we are going to make it" Rifkin says. We have the technology as a "nervous system" to think universally as a family, like it happens in nature disasters for example as earthquakes hit a region and a lot of solidarity helps to relief from that. Think of Haiti, or those earthquakes in Italy in the last decades. Twitter, Youtube, facebook, internet etc. are forming this nervous system in a kind of evolutionary process with all this frequently going astray in a trial and error evolutionary mode. "We have to begin thinking as an extended family, we have to extend our identities to think of the human race as fellow sojourners and to extend this to other creatures and the biosphere" Rifkin says. This requires to fade the tribal, religious, ideological, nationalist identities more and more in the background.

What if we encounter Artificial Intelligence with probably no empathy in "its genes". With no frailties and acknowledgement of death in its "consciousness"? How will we coexist—humans with their empathetic abilities as well as their abilities for narcissm, materialism, violence and aggression—with Artificial Intelligences in a "deep learning" mode? Will they once collide on a global scale or will we learn to coexist on a small scale in our organizations   as hybrid intelligences hopefully with empathetic humans in the driver's seat. "We won't have the time of thousands of years like mankind had for the developments before" as Rifkin has described so vivid and clear. So we are urged to find ways to coexist on a small scale to be able to learn fast before the "singularity" as Ray Kurzweil predicts will hit us. So let us try to view our organizations and organizational units metaphorically as "hybrid intelligencies" in order to learn as organizations and as society. Let us go astray sometimes and develop as a learning organization with human empathy in the driver's seat as "top level controller". This will help us to learn as a society and as mankind in order to preserve our planet and the biosphere—and mankind itself.

6.2.13 How Is Collaboration Between Man and Machine Expected to Develop?

Let us step back from these more philosophical views a bit and look at future manufacturing industry:

It's not about human versus machine, it's about human + machine we can read in books, articles and blogs. The winning equation for successful factories will be combining advanced AI with new business processes and models.

For decades, employees and machines have largely worked independently throughout the production process. Industrial robots were simply too dangerous to be near humans. Now, advances in AI and robotics are opening new possibilities for intelligent and efficient human-machine collaboration on factory floors.

Manufacturing robots are finally becoming more flexible, approachable coworkers who are safer to work with—compared with the robots of the first generation who worked segregated from humans due to security aspects—like fenced in welding robots in automotive industry.

But there are some challenges for such closer collaboration. On the one hand we need a more segmented and quality-differentiated net for security sensitive applications like manufacturing systems. Imagine a hacked and manipulated manufacturing unit and how much damage criminal energy would be able to cause. On the other hand we need filters for relevance close to the process requiring intelligent sensors being part of a hybrid intelligence in manufacturing with distributed but coordinated intelligence. Imagine the sensor data overflow without decentralized intelligent filtering for relevance. Industry 4.0 as concept for advanced industrial processes will only succeed, if we manage to create robust hybrid intelligence in collaboration man+machine. The human parts of the hybrid intelligence will be highly knowledgeable employees, who have to be aware of the digital as well as the overall strategy of the organization and who have the ability and the means to search for and acquire the knowledge they need to accomplish their complex tasks in collaboration with AI elements of the hybrid intelligence on the one side and to keep AI and its learning algorithms under control on the other side.

The only way to manage this complexity will be robust platforms that hide the complexity as well as provide the necessary security and safety of the system levels beneath though providing the ability to analyze and interfere if necessary for the supervisors of the systems who themselves are supported by AI enabled control facilities.

Following an article on Singularity hub (Berman 2016), John Hagel, the co-chairman of Deloitte's Center for the Edge referred to the decades-long economic trend Deloitte calls the "big shift". Hagel said the economy is splitting into two parts—products and services are fragmenting into smaller and smaller pieces, while the platforms knitting them together are getting bigger. The need for organizational agility—enabled by platforms—is one factor that will force large organizations to become more innovative in both their operations, management, and strategy.

On the other hand Big Data and AI help to make processes transparent and better manageable thus securing that it can be observed what is happening in hybrid intelligencies.

Sandy Pentland, director of the Human Dynamics Lab at the MIT Media Lab has been leading the big data revolution for decades. Pentland's work is helping to build a society enabled by big data and showing how to scientifically understand human interaction (Berman 2016). For example he showed how big data can help large companies better understand the behavior and communication of their employees and eliminate outdated business practices.

The scientific discipline of Organizational Behavior—how people behave in an organization and how to influence them—was established as humans were the only intelligent beings in an organization, forming and managing the organization, doing work together with machines and IT, that supported them to automate and accelerate business processes.

As now AI is on the playing field—together with humans (who mostly did not create it and do rarely understand what it is doing)—organizational behavior is about to change too. To be precise, we have to consider the behavior of humans in an organization as well as the behavior of an organization or an organizational unit as a whole. This behavior could be affected by AI profoundly—especially in case we just let AI "happen" in our organizations. The subconscious mind of organizations has to be deliberately shaped as described in chap. 5.

Let us start with "narrow AI's". Every time you use a search engine you're taking advantage of data collected by 'smart' algorithms. When you call the bank and talk to an automated voice you are probably talking to an AI…just a very annoying one. Our world is full of these limited AI programs which we classify as "weak" or "narrow" or "applied" AI.

Saenz (2010) says that these programs are far from the sentient, love-seeking, angst-ridden Artificial Intelligences we see in science fiction, but that's temporary. "All these narrow AIs are like the amino acids in the primordial ooze of the Earth. The ingredients for true human-like Artificial Intelligence are being built every day, and it may not take long before we see the results", Saenz says.

How did we create the jungle of AI that surrounds us today? Saenz writes that Rodney Brooks of MIT (also one of the founders of iRobot) took a new approach to AI. Instead of developing AI from the top-down he looked at building things from the bottom up. Instead of artificial reasoning, he looked at artificial behavior.

The result were robots that based their actions upon basic instincts and patterns. "iRobot's Roomba doesn't vacuum a floor with high-level reasoning about how the carpet should eventually look, it performs a bunch of different cleaning patterns until it knows the whole carpet's dirt-free." Saenz writes.

That's behavior-based AI, and it's quite powerful. Along with increased processing power, Artificial Intelligence really took off in the 90s. Using modular and hierarchical techniques like Brooks' behavior-based approach, researchers were able to create a variety of AIs that did things. These weren't philosopher programs, they "worked for a living". Data mining, inventory tracking and ordering, image processing—these jobs all started falling to AIs that built simple patterns into algorithms that could handle dynamic tasks.

Now that list of tasks has expanded. We're slowly building a library of narrow AI talents that are becoming more impressive. Speech recognition and processing allow computers to convert sounds to text with greater accuracy. Google is using AI to caption millions of videos on YouTube.

Likewise, computer vision is improving, so that programs can recognize objects, classify them, and understand how they move. Many machines might even be dangerous because they simply lack self-awareness and the ability to understand their surroundings.

This is why, up until recently, robots haven't been able to work side-by-side with people in industrial settings. Now robots are available, that can work with people for example Baxter by Rethink Robotics or telepresence robots by iRobots. Baxter for example has eyes on a screen to show where the machine is "looking" to signal its next task. Narrow AI isn't just getting better at processing its environment, it's also understanding the difference between what a human says and what a human wants.

Thinking along this development trajectory we can imagine that complex systems beyond manageability may emerge and on the other hand we can imagine that AIs emulate "empathetic behavior". Paro, the seal robot, is used for therapy of patients suffering from dementia at certain clinics already. The efficacy of the robot therapy was evaluated, which detects patient's cortical neuron activity from a 21-channel EEG. The results from preliminary experiments show that robot therapy has a high potential to improve the condition of brain activity in patients suffering from dementia (Wada 2008).

Today Paro is widely used as *medical commitment robot* and provides help to people with dementia. The robot has under his fluffy bright fur tactile sensors and can perceive it, when a human caresses him. Paro reacts on it with the movement of the tail and the head and the eyes. The robot also responds to sounds and can learn a name. Paro makes even sounds similar to those of real harp seal cubs. Unlike these Paro is programmed so that it is not at night, but only on the day active.

Though this is a simple example of emulated empathy (not going deep!), we increasingly hear about chatbots on service helplines which are hard to distinguish from human responders on a hotline or a helpline. But will such emulated empathy ever be able to influence automated decision making by AI? Or will it be simply a tool or an element for conviviality in hybrid intelligence?

Will we be able to create a sustainable future in this world of exponentially developing technology or will we surrender? How can we keep ourselves as humans in a position of governance? Where can we learn best? Maybe from building and experiencing hybrid intelligences in our organizations—starting small and growing the AI part in a controlled way—keeping the well being of us as human beings and as society in mind.

6.2.14 Creating Sustainable and Equitable Futures in a World Increasingly Dominated by Technology

In 2016 the US National academy published a paper on a teamwork concerning frontier collaborations of art, design and science, engineering and medicine and their ideation, translation and realization (NAP 2016): The team was asked to tackle the issue of creating sustainable and equitable futures in a world increasingly dominated by technology. The team decided that the fundamental issue pertaining to this larger problem is a pervasive lack of empathy within the global community and an overwhelming expansion of feelings of indifference. Therefore, they proposed the following project: **Empathy in an Expanded Field: Stemming the Rising Tide of Indifference**.

One potential manifestation of this program is the creation of avatars designed by those who have respiratory and other diseases that are made worse by poor air quality. These avatars would be available to neighbors, the community and the world through a cell phone application that would allow people to see who is truly affected by their decision to drive to the shopping mall every day, when doing so would further debilitate someone suffering from a condition such as asthma. Developers would create the application, but then users would have to download it and log into track how their neighbors were doing on a given day.

The suggestions can be summoned as follows presenting the solutions proposed by the different teams:

Create an Open Data culture

Open data means providing unrestricted data to everyone. There is a lot of data within the public sector and within science that are not as useful to benefit society as they could be if they were available to anyone. As one example, if we wish to understand the elements of the integrated system that is the Earth's atmosphere, oceans, and biosphere, the way those elements interact and how they have changed with time, it is necessary to be able to collect and analyze environmental data from all parts of the world and from many sources. In the public sector in recent years, many state and local governments have put effort into open data projects that would inspire developers to create apps and find ways to use public data to bring value to their communities.

Foster Innovation, Creativity, and Action

Herbert Simon once famously stated: "design like science is a tool for understanding as well as for acting." What might be examples of principles, methods, processes, and strategies of a successful art/design and science integration that may accelerate our ability to both innovate and act?

Consequently, the team developed a "virtual Besso," a creativity tool, named after the assistant of Einstein who served as partner and sounding board to the genius on long walks discussing Einstein's ideas. The goal of a creativity tool is to accelerate the birth of novel ideas. The team thinks that a creativity tool should also provide emotional support and encouragement, facilitate divergent thinking, and foster collaboration among individuals. Virtual Besso would be a program that provides users with a creativity-enhancing companion and sounding board. The same way that Einstein and the real Besso walked together, virtual Besso, an augmented reality program, would provide creative guidance to a user while they walk. Using augmented reality glasses, Besso would offer users visual and auditory stimulation that might enhance their creativity. On a walk, a user would be able to converse with Besso. A user's conversation with Besso would prompt the appearance of visual stimulation via augmented reality glasses that would enhance a user's immediate surroundings with graphics superimposed on the natural environment. These graphics would be linked with elements in a user's environment, synched in a way that integrates the augmented images into a user's immediate surroundings.

Users could then follow an associative trail of ideas away from their original search terms.

Users would be able to share their storyboards with other users. The process of sharing storyboards is meant to encourage collaboration between users. The team envisions Bessos across the world linking together according to their users' common or complementary interests to make connections between users. Bessos would introduce users in the hopes of spurring collaboration. The team envisions users' Bessos forming a worldwide network or "invisible college" that brings together creative thinkers from many different fields.

Harnessing the World Brain to Address Urgent, Global Issues

Tools like virtual Besso would help "Citizen science" or "crowd-sourced" science to become increasingly important as a valuable means of harnessing the "world brain" to address critical social, ecological, or cultural issues. "Harnessing the World Brain to Address Urgent, Global Issues" requires access to the minds of billions of people worldwide in an attempt to provide human intelligence to solve tasks that might be too complicated even for the largest computers. This network of brainpower is the global brain, a concept that serves as the namesake for the proposed smartphone app called G-Brain.

G-Brain, like other apps, would pay people modest amounts of money to contribute their time to different types of projects—some of which could help scientific advancement. The app's ability to access data from the input of billions of people could be analyzed to develop information that might prove useful when addressing urgent global issues. However, the app is not unlike other platforms that already exist. Some tasks with G-Brain would require instant, live interaction between task requesters and workers. For example, a team member suggested that G-Brain could be useful in emergencies where a crisis overwhelms healthcare providers, thus affecting their ability to triage patients effectively. In such a scenario, G-Brain workers would provide instant feedback to help healthcare providers assign different degrees of urgency to patients and decide which patients require immediate assistance, the NAP study describes.

By exploring scientific unknowns through the vehicle of artistic collaboration, the team proposed that through a Web-based platform for connecting artists and scientists, the community could be compelled and inspired to do the same. As an artist stated during a meeting, "As an artist, I can take a big problem and represent it in a form that people can interact with and relate to…. I'm not [necessarily] trying to propose a solution … but to make the problem more visible." In areas where scientists may seek to eliminate variables and find the simplest explanation for data, artists may be able to expand the question beyond the confines of the experiment. By combining art and science perspectives to encourage cross-pollination of thought, the potential to enhance both scientific thinking and artistic expression is unlocked and the unseen can be visualized. Think of the views of the poet and author Clemens Setz as cited in the sections above (e.g. sect. 3.2.15).

Creating human-centered cultures with human-centered technologies

Too often, rather than broadening people's viewpoints, modern technology reflects people's own opinions, views, and biases back to them, an effect known as the echo chamber. The team wanted to use technology to decrease isolation, broaden people's views, and promote empathetic responses to opposing viewpoints. In effect, the team proposed to use technology to allow people to escape the echo chamber.

By creating an environment in which people encounter different viewpoints less frequently, the echo chamber can lead to decreased empathy. The team thinks this is a significant social problem because decreased empathy can inhibit people's ability to collaborate and solve complex problems, whether locally, nationally, or globally.

The group considered what promoting empathy might look like. One idea is to create an experience that would allow someone to step into someone else's skin and feel what someone else is feeling. Could such an experience give someone a better understanding of why different individuals have different points of view? Could it help develop greater empathy for others? Would it ultimately teach people more about themselves?

For example tram drivers in Graz, Austria went through a program to experience how elderly people feel-simulated with weights, glasses emulating reduced sight, appliances reducing mobility of joints etc. The goal was to increase patience of the tram drivers with elderly passengers.

The installation, which the team plans to call the Asymmetric Mirror, would reflect an alternate version of the participants with the goal of opening their eyes to the viewpoints and experiences of other people. The team decided the Asymmetric Mirror would involve three experiences, sight, touch, and sound, which would use existing technology as well as technology in development to engage multiple senses while exploring how technology can be harnessed to induce an empathetic experience (NAP 2016).

Though the challenges for the teams were how to benefit the society as a whole it also includes ideas that would be helpful in creating hybrid intelligences in organizations with tools becoming part of their subconscious mind. To bring art, architecture, design together with the interests of organizations and their missions and goals to contribute to the commons and not only to contribute to the individual benefits or the benefits of the shareholders is a valuable idea to follow for organizations. For example the ars electronica center in Linz, Austria with its regular work and the always state of the art exhibitions as well as its yearly festival serves the public and is used by companies as a framework for thinking about future challenges, products, services and ideas (Ars Electronica Center 2016).

Such activities are intended to initialize and support "Start ups for the commons" and not only "Start ups for profit" though these are addressed too. Organizations are well advised to use such facilities in order to equip for the future and to create hybrid intelligencies that help them in their daily as well as their strategic decision making in order to support their missions and goals.

We have had now extensive considerations about how hybrid intelligences can be formed and shaped, when teaming man and machine and we have discussed decision making as the core activity in organizations in general as well as in hybrid intelligences from different perspectives, in particular

- self-tuning organizations,
- management structure,
- Real life gaming as "story telling" in hybrid intelligencies,
- blending analytics with intuition,
- reorganizing around decision making,
- the drivers of business performance,
- required managerial skills and
- team conditions' impact on organizational behavior.

Then we considered the aspects of conviviality and empathy as guiding principles when creating, governing and managing hybrid intelligencies. We then tried to fathom how collaboration will develop in hybrid intelligences forcing man and machine in close collaboration, starting with manufacturing where robots and man will collaborate to general aspects of organizational agility, to behavior based AI until robots emulating empathy. Finally we considered practical ideas for creating sustainable futures dominated by technology, thus stemming against the tide of indifference by exploring potentials for expanding empathy and creativity from the Virtual Besso to the Assymetric Mirror to harnessing the "world's brain" by creating human-centered technologies and human centered cultures in our organizations with their future hybrid intelligencies. The integration of art and artists in this developments seems very promising.

Let us—as last chapter before looking for rules and ethics for hybrid intelligencies —have a look at deep learning as a core technology of AI, the technology attributed with a tremendous potential for incredible impact—for the good and the bad.

6.2.15 Will Deep Learning Enable AI'S to Have Emotions?

Now let us have a look on one of the key features of AI—deep learning—on the formation of hybrid intelligencies and the relationships between humans (in a hybrid intelligence) and the AI parts of it.

Bidshahri in his essay on "How AI will redefine love" (Bidshahri 2016) says, that advancements in deep learning have allowed computers to rival humans in many areas where they've traditionally struggled, such as pattern recognition, natural language processing and computer vision. The beauty of deep learning is that these artificial neural networks train themselves in such tasks, allowing them to improve their skills without human intervention. Many projects are also attempting to move AI beyond automated visual and language tasks. "One secretive AI firm, Vicarious, is striving to teach computers imagination, and Google is programming them to be creative", Bidshari reports.

The AI in future may seem to have emotions, memories, a sense of continuity, a capacity to self-reflect and the ability to use language to communicate all of this. But how will we know if an AI has the true inner experience of consciousness or if it has simply been programmed to create the illusion of consciousness? So we are deep in the metaphoric and analogy view, where we partly derived the concept of the subconscious mind of organizations from.

Known as the hard problem of consciousness, this remains one of the biggest mysteries in neuroscience as scientists attempt to explain how we have phenomenal subjective experiences—also known as "qualia". Badshiri reports, that Princeton neuroscientist Michael Gazzaniga says, "I don't know if you're conscious. You don't know if I'm conscious. But we have a gut kind of certainty about it. That is because an assumption of consciousness is an attribution, a social attribution."

The same social attribution can be applied to AI. The line between the consciousness experienced by intelligent machines and human beings might be blurrier than we would like to admit. Our brains are the biological source of all our emotions, memories and subjective experiences. Given that fact, theoretically, many experts believe that if we were to replicate the structure and function of the brain, we should be able to replicate all the experiences that come with it. "That also means replicating the emotions and subjective experiences", Bidshari writes.

Obviously we are—with "The line between the consciousness experienced by intelligent machines and human beings will be blurrier than we would like to admit"—on the way to human and Artificial Intelligence meshing so profoundly that this really might be considered "hybrid" in a very deep manner—beyond what we described above. Remember the definition of hybrid in biology and have in mind the tremendous developments in Computer-brain interfaces with the vision presented by Facebook-manager Regina Dugan on the Developer-conference F8 in San Jose in April 2017 that without implants they expect to enable humans to write text just by "thinking it". 60 experts are working on that. Imagine: Will Facebook be reading our minds soon? Is that going to develop to a nightmare? Or should we trust Elon Musk cofounder of PayPal, founder of Tesla and big shareholder of Neuralink, a company that is too working on interfacing our brain with digital enhancements. Elon Musk obviously believes that this would help humans to enhance their intelligence and thus enable him to prevent a takeover of Artificial Intelligence, something Elon Musk, together with other prominent leaders has been warning of in several statements. So the Hybrids from mythology (as mentioned in the beginning of Sect. 6.2.) would turn out to be "hybrid intelligencies" in a very immersive form with an incredible high degree of integration of human and Artificial Intelligence—like Cyborgs from Science fiction.

Eminent sci-fi author Arthur C. Clark writes in *Odyssey Two*: "Whether we are based on carbon or on silicon makes no fundamental difference; we should each be treated with appropriate respect." We can imagine a time when individuals start fighting for the right to legally and socially declare their love for AI. Consider how society has evolved to accept more of different forms of relationships over time.

Our technology, powered by Moore's law, is growing at a staggering rate—intelligent devices are becoming more and more integrated to our lives. Futurist Ray

Kurzweil predicts that we will have AI at a human level by 2029, and it will be a billion times more capable than humans by the 2040s (Kurzweil 2005). Many predict that one day we will merge with powerful machines, and we ourselves may become artificially intelligent.

"In such a world, where our own existence will be largely non-biological, it is only inevitable that we eventually accept being in love with entirely non-biological beings", Badshiri argues.

Why love at all? The capacity to feel passionate and powerful forms of love could be amongst the many evolutionary advantages that have allowed humans to progress so far as a species. It stimulates the drive to procreate, stay alive and keep one's loved ones alive. "AI may ultimately be able to soothe the human condition and relieve us of the existential angst of loneliness by granting access to something we all crave—the powerful desire to love and be loved", Badshiri closes.

After these far reaching ideas and concepts to what AI might evolve and to what dominant role AI might develop in hybrid intelligencies—emulating deeply human skills and abilities like intuition and love—we may doubt that we might be able to stay as humans in the driver's seat. But at least it is us as humans who create and shape those hybrid intelligencies in our organizations. So let us be optimistic and look how we can best organize around our key activity in our organizations—in decision making together with our "AI-companions" (as elaborated in the chapters above). The human species will not stop developing those technologies because we consider them dangerous and "transhuman". We only can try to mitigate the risks and openly discuss the dangers and problems as society—but not only on a national level but globally.

Given the tremendous capabilities which deep learning from Big Data—often provided by narrow AI's themselves—will provide (as described in several sections before), and given that some form of "General AI" will supposedly be much more centralized (like the search engines today or Watson winning jeopardy) it is clear: We will need new rules and ethics to cope with all this variety of AI's. Not only from the global and societal view but also rules and ethics applicable in everyday life with AI's —in our organizations with their hybrid intelligences stumbling along the bumpy road through a strange new landscape called digital transformation being hardly capable to distinguish what is real, what is virtual and what is fictitious, trying to take the best possible decisions. Organizations are already influenced and partly controlled by their subconscious minds—soon with AI and deep learning incorporated in it.

6.3 Rules and Ethics for Hybrid Intelligences and Digital Transformation

From the viewpoint that what is technically possible will eventually find its application, we should think about how we can make arrangements in the organizations so that Artificial Intelligence, automated decisions and self-learning and self-optimizing systems, algorithms and processes are used wisely. These issues are broadly discussed on the societal level in numerous books, journals, conferences, talk shows etc. Futurologists with their shows (not keynotes) are telling their audiences what might come and that it is basically inevitable. But they rarely offer actionable proposals how to deal with this future scenarios exept that we have to be adaptive. Organizations are decisive building blocks of our society. So organizations of all kinds (profit, non-profit, public or private, associations etc.) will be those who have to deal with these new technologies early, organizations will serve as test beds for the digital transformation of our society and will have to be the first to find rules and ethics to deal with it. At least they will arise questions and thus induce new rules in our laws, as we realize that we will need modified legislation in order to manage hybrid intelligencies in the societal context. Basically it is therefore to ensure

- that organizations manage that processes are implemented for the evaluation of learning progress when continually improving,
- that possible negative and positive side effects are observed closely,
- that it is ensured to perform a conscious reflection before implementation and continuation of the innovation, if the observed changes are not plausible.

This is all the more urgent, as our methods of generating information and interacting with information are changing. The rules by which we govern, and the values that we want to protect with the rules should therefore be adapted.

Consider the following scenario: *I drive with a fully automated self-driving car and read the newspaper in the car. From left a child jumps on the street, from right a truck approaches at high speed. The machine has to decide whether it—as the driver—saves me and the child gets hurt or dies or whether it is driving the car into the truck. How will the machine decide? Probably the systems would protect the driver and sacrifice the child, although the driver's survival would also be possible in the collision with the truck.*

Who is responsible for this decision? The driver? The carmaker? The creator of the algorithm? There are a number of ethical and legal issues that we cannot leave alone to the technologies.

Is this kind of car already an electronic or artificial personality in terms of an adopted new legislation? With which rights and opligations?

Another scenario: *During a search algorithms analyze networks of people and elicit backers of a criminal organization. The probability for the accuracy, however, is given only to 95%. What if the remaining 5% hit me?*

Another scenario *is the electronic profiling in loan applications and the data based assessment of the creditworthiness of a person. Knowing that it is in statistics always "only to probabilities", we know about possible miscalculations, incorrect data and data interpreted wrong or "fuzzy". Thus existences can be affected massively.*

"We need some sort of a digital ethos" Hofstetter says (Hofstetter 2014). "If large corporations use Artificial Intelligence unhindered and unlimited, there is the danger of new dictatorships." The same may apply for democratically not controlled (government) agencies etc.

The sociologist Niklas Luhmann wrote in Theory of Society (Luhmann 2012): "The main function of memory is forgetting, thus preventing the self-locking of the system by clotting of the results of earlier observations". Forgetting frees capacity for attention and communication. Social behavior is therefore only possible by forgetting. Our brain stores sensory impressions initially for a very short time and then they are rated. Only what is considered "important" will pass in our long-term memory. So let us consider issues like unlearning, decluttering etc. in order to be able to develop.

Before adressing particular rules for the digital transformation and the handling of hybrid intelligencies let us have a look on an analogy that may help us handling these complex issues.

Samuel Arbesman, a complexity scientist and writer, published the book "Overcomplicated: Technology at the Limits of Comprehension" (Arbesman 2016). It's a guide for dealing with technologies that elude our full understanding. In his book, Arbesman writes we're entering the entanglement age, a phrase coined by Danny Hillis, "in which we are building systems that can't be grasped in their totality or held in the mind of a single person." In the case of driverless cars, machine learning systems build their own algorithms to teach themselves—and in this process they become too complex to reverse engineer. And it's not just software that has become unknowable to individual experts, says Arbesman.

"Machines like particle accelerators and Boeing airplanes have millions of individual parts and miles of internal wiring. Even a "technology" like the U.S. Constitution, which began as an elegantly simple operating system, has grown to include a collection of federal laws 22 million words long with 80,000 connections between one section and another", Arbesman writes (Arbesman 2016).

Arbesman makes a compelling case that we will need a natural science approach. Consider, for a moment, how we deal with weather. "While we can't actually control the weather or understand it in all its nonlinear details, we can predict it reasonably well, adapt to it, and even prepare for it." Arbesman suggests we may soon model computer glitches the same way. "We'll need interpreters of what's going on in these systems, a bit like TV meteorologists," he writes. In the entanglement age, our relationship to technology has evolved to be like our relationship to nature. We're returning to the jungle—only this time we've constructed it ourselves.

Though this is an interesting view of how to approach complex systems, we will need more than a "TV meterologist" to navigate safely through this jungle, the digital transformation confronts us with. We need a set of adapted rules that have to be incorporated in the legislation. I would therefore make some suggestions (partly derived from literature) for dealing with data innovatively and implementing Artificial Intelligence in our organizations. Much of it also seems valid for how our society is dealing with these technologies:

First, on the subject of data sharing and data protection (partly following (Mayer-Schönberger and Cukier 2013) and their proposals for dealing with Big Data—mingled and extended with other ideas):

- We have to move from the "Data Privacy" perspective to the perspective of the "Responsibility of Data Users".
- For this we need to get away from the formalized rituals of the Privacy Policy with its signed informed consent of the affected individual. This often seems anyway only an illusion of self-determination and control for the persons concerned.
- Rough use categories should be regulated in privacy laws—it has to be clarified if limited standardized protective measures should be defined or not.
- Binding rules for the assessment of a planned use of data should be adopted as well as for precautions to be taken.
- Organizations should have to conduct a formal investigation for any purpose where they use personal data, considering the potential impact for those affected and document it (see the scenario about creditworthiness).
- The data user is legally liable for what he does with the data. He has to perform self checking. So there is no obligation connected for the organization to disclose its trade secrets. In case of accusations or in routine external audits the access to sensible personal data have to be opened for sample data and the purpose of the access has to be clarified to the authorities or auditors.
- An applicant has—on request—a right to be told the reason and which data have been used.
- Concerning the "specter of eternal memory" lawmakers should set maximum storage periods for data. This would also encourage the early data reuse. Alternatively or additionally, should data be 'disarmed' by being made anonymous—at least after a certain time. This is usually sufficient for the beneficial use in long-term considerations as big data.
- Big Data analysis should be granted only limited probative value in court. This is a kind of "prevention" against illegal use in forensic environment.

Thus, a social consensus should be found, that people will continue to be judged according to their responsibilities and their actual behavior and not by "objective data analysis" and "profiling".

For prediction, machine learning and robotics, also rules should be given:

- For the responsibility concerning privacy the above formulated rules should be valid.
- Algorithms and Data sources for prediction as well as deep learning must be disclosed, in order to get eventually certified (depending on the risk for misuse and severe faults that might accompany the use of these algorithms), similar to as it is handled with medical devices today.
- Certification (compulsory in sensitive areas of use—to be defined): Independent experts should declare the algorithms for the respective use being appropriate (like with medical devices etc.).
- It has to be evaluated if "artificial personalities" should become part of legislation.
- AI components must have a veto mechanism combined with a comprehensive description of the AI content and algorithm.

- The dominance of a single "data baron" is to be controlled—if necessary by antitrust laws. Monopolies are to be prevented if possible.

The futurist and science fiction author Isaac Asimov formulated laws for robotics in 1942 (Asimov 1950) that are impressive in their simplicity:

1. A robot may not injure a human being or, through inaction, allow a human being to come to harm.
2. A robot must obey the orders given to it by human beings except where such orders would conflict with the First Law.
3. A robot must protect its own existence as long as such protection does not conflict with the First or Second Laws.

For the actual programming of industrial and domestic robots today, different security standards apply. It starts at EU level with the Machinery Directive. The transposition into national law takes place for in the respective country like in Germany by the Machinery Directive. In Austria this is regulated by the Machine Safety Regulation. As with other regulations this is very troublesome for entre-preneurs. The problem of fuzziness and quality of software, algorithms and data is evident. There are authorities and instances to be established, that should not hinder the innovation, but ensure that a minimum level of security, social and societal acceptability and sustainability are provided. It is to be clarified who reviews algorithms concerning their security or harmlessness, and who should be allowed to certify. Ultimately, it must be clarified by law who takes the responsibility, who takes the risk and who is liable (see the first set of rules above). These laws should be internationally accorded (I know that sounds like fantasy having the actual situation concerning taxation, CETA and TTIP in mind).

There already are numerous regulated areas subject to certification in the economy in the interests of the security of people. For example, the FDA (Federal Drug Administration, USA) has approved telemedicine robots that move autono-mously among people in the corridors of a hospital.

These mechanisms and tools for the protection of a society and civilization also have to be introduced in the partly virtual world of the algorithms. Here regulatory policy still has much to do. The companies urgently must agree on a basic ethical consensus on these issues. The pace of innovation in these fields is currently enormous.

Let us take a specific and detailed view exemplifying the challenges ahead: Consider—in terms of brain research and the attempted re-engineering of the brain —(Fig. 3.1) the mechanisms of Neuromorphic Computing and the PRTM-model (pattern recognition theory of mind) by Ray Kurzweil, where the transmission of signals is triggered by actually achieving an analog threshold. Thus in pattern recognition, even missing parts of the pattern can be compensated in the brain and lead to combinations of such transmissions of signals that enable that the pattern can be recognized in spite of the missing lines.

Now let us take the "helicopter view" again: So one thing seems to become evident: the 0/1 or yes/no world of digitization, which has given us such techno-logical advances, is probably only a transitional phase for the next evolutionary step. Though the "0/1 world" will probably remain as tool on lower platforms of the entire systems as they serve and probably will partly control us. But we do not know how this next evolutionary step will look like. Will it be the singularity? Will it be some form of transhumanism? There are huge uncertainties in spite and because of these exponential developments in technology. Developments not only in ICT as described in this book but as well in synthetic biology, nanotechnology etc.—all those exponential developments potentially reinforcing themselves mutually.

We are now mostly living in an analog world. As humans we are used to this sort of world. At the end of Sect. 4.7. we have argued that "The analogue will become the new Bio" and maybe the rich people will afford interaction with humans and make holiday on a real beach wheras the poor people will spend their holiday at home but travel in virtual reality and—in case of sickness—being cared for by a robot because they cannot afford interaction with human doctors and nurses.

Maybe "Digital" in the end will have been a tool for a transitional phase, preparing the merging of biology and electronics in "real hybrids" (with a brain-computer interface)—to "next generation" hybrid intelligences. Today we already experience "analog" (perceived via our senses) expansion with these new technologies and opportunities—even if these extensions may be virtual: augmented reality, expanded perception etc.

The challenge will be "to go down this path" in an "evolutionary mode". First we have to walk the path to the first generation hybrid intelligences in a very controlled manner—how to incorporate them in our organizations thus modifying parts of our organizations. This is today's challenge.

In parallel and overlapping we'll thus gradually develop "intelligent organizations" as hybrid intelligencies, in which more and more already runs in the "sub conscious mind of organizations" in the depicted cascaded structure (see Fig. 4.3) within the meaning of "thinking fast" by Kahneman (System 1). This subconscious mind will hopefully be consciously designed and automation will not only "creep" in, but will be considered and installed in a controlled way and will be regularly evaluated—at least in successful and sustainable organizations.

Let us shift the focus beyond the organization itself and its interactions. Let us view our society as a whole: the interaction between individuals, between individuals and organizations as well as between organizations. One technology I'd like to introduce at the end, a technology that is currently in its hype phase and that has tremendous potentials (following those who propagate this) but also tremendous risks is the blockchain technology. It has a lot to do with rules and ethics when extending this view to hybrid intelligencies working across organizations.

The first generation of the digital revolution brought us the Internet of information. The second generation—powered by blockchain technology—is bringing us the Internet of value: a new, distributed platform that can help us reshape the world of business and transform the old order of human affairs for the better (Tapscott and Tapscott 2016).

Blockchain is the ingeniously simple protocol that allows transactions to be simultaneously anonymous and secure by maintaining a tamperproof public ledger of value. Though it's the technology that drives bitcoin and other digital currencies, the underlying framework has the potential to go far beyond these and record virtually everything of value to humankind, from birth and death certificates to insurance claims, to financial transactions, smart contracts and even votes. This technology is public, encrypted, and readily available for anyone to use. It's already being wide-spread adopted in a number of areas. For example, many of the world's biggest financial institutions, including Goldman Sachs, JPMorgan Chase, and Credit Suisse, have formed a consortium to investigate the blockchain for speedier and more secure transactions (Tapscott and Tapscott 2016).

This technology promises cross border money transfer like secure eMail—without intermediaries like banks and clearing houses. Figure out "smart contracts" where the rules of a transactions are laid down and the contract is executed automatically—step by step—with the fulfillment of a promise e.g. the delivering of a product promised being one step that is automatically followed by the action of the partner of the contract as layed down in the digital contract—for example the payment. This huge ledger behind where all steps performed are documented is said to be immuteable and unhackable. For example when you want to know how many bitcoins you have (on your virtual account) this is generated from the relevant transactions documented in this giant distributed ledger (each transaction being a block added to the chain with its reference to previous blocks). This principle and the way e.g. how bitcoins are generated ("bitcoin mining") needs tremendous ressources of computing power and energy input. The promises of blockchain technology are: protecting rights through immutable records, ending the remittance rip-off, enabling a sharing economy, boosting productivity in financial services (as far as employed people are concerned) etc. This (according to Tapscott and Tapscott 2016) should be leading to an "extended enterprise", further on to a "business web" and finally "distributed value creation" requiring some sort of "networked intelligence". It is possible to eliminate intermediaries and unbundle the power of platforms like with Uber, Amazon etc. with their generating of profits without value adding to a product or a service. Thus they leave the service provider to himself (e.g. the driver with his car at Uber). These drivers are exposed to full competition without protection against dumping etc. forcing them to incredible flexibility and availability in case he has to live from that. In addition this technology promises to be simultaneously anonymous and secure—seeming to be a very seducing promise. It also promises for example to enable secure e-voting.

As with major paradigm shifts that preceded it, the blockchain will create winners and losers. And while opportunities abound, the risks of disruption and dislocation must not be ignored. Potential show stoppers according to Tapscott and Tapscott (2016) are: Technology is not ready for broad use, the energy consumed is unsustainable high, governments will stiffle or twist it because blockchain is a job killer, powerful incumbents of old paradigms will usurp it (think of the consortium of financial industry) etc.

The promise of anonymity is more of a threat than a promise. We know that bitcoin as cryptocurrency is a popular currency in the darknet and the financing of criminal action. There are companies who aledy have bitcoins available to be able to pay in case of a ransomware attack and blackmailing. It would be better to have a hard regime concerning privacy with high penalties in case of fraud and misuse and combine it with banning anonymity forcing all participants to authentification, so every user can see who is the individual behind and who is the responsible and liable manager of a company you are dealing with.

In case you miss anonymity: There are other possibilities to practice anonymity in case of reporting corruption etc. for example by using advocates who guarantee anonymity and can give report back to the whistle-blower. Imagine full-blown blockchain infrastructure and you do not get access to your bitcoin account or your smart contract because your password has been stolen. Or your digital identity as a whole has been stolen and/or altered. Imagine this in a world of smart contracts where the intermediaries have dissappeared due to the productivity promises given by blockchain technology—no notary when transfering money in exchange to the land register entry when you buy a house etc.

The anonymity in (secure) internet and blockchain technology, with its surrounding potentials of misuse, identity theft etc., with its potential to eliminate persons and institutions of trust like a notary, a bank, members of your election commission when voting etc. does not provide a sense of trust. We are thus on the way to loose trust to people and to institutions. Liability will fade away. Some will win, many will loose. Society will loose. Trust is the decisive factor in a community be it small or large. We are going to erode the fundamentals constituting our social cohabitation. When you look at the actual political processes all over the world we see alarming signs of de-solidarisation, loss of trust, democracy in danger and so on. Artificial Intelligence in combination with blockchain technology and anonymity giving room to criminal action, postings of hatred, mass-postings by AI's enforcing "echo-chambers", manipulation by chatbots etc. endanger social cohesion massively. Regulation, ethics and political leaders under democratic control are massively challenged.

There won't be plenty of time to enforce suitable regulations before the exponential technologies form something in favor of a few big players who will surely promise us to be philanthropic and idealistic and support the commons-thus forming new feudalistic structures.

But let us be confident: Humans will still be sitting in the "driver's seat". But we have to fight for it. If we do it right, organizations will be populated with hybrid intelligences that follow ethic and legal rules and enable our organizations to be purpose-led, self-managed and networked and to reach their goals in the ever changing world of digital transformation and disruption. But in the end for the benefit of the commons, for the benefit and development of society.

Basically, the following principle should be ensured for people, for their relationship within and to the structures, organizations and in particular the use of new technologies:

People do not serve the structures but the organizations and their infrastructures serve the people.

Literature

Aguirre D, von Post R, Alpern M (2013) Culture's role in enabling organizational change. A 2013 Booz & Company study

Amit R, Zott C (2012) Creating value through business model innovation. MIT Sloan Manag Rev, Magazine: Spring 2012—Research Feature 20 Mar 2012

Arbesman S (2016) Overcomplicated: technology at the Limits of Comprehension. Penguin Random House, New York

Ars Electronica Center (2016) Center, exhibitions, festival, futurelab, solutions. http://www.aec.at/news/

Ashkenas R (2011) Reorganizing? Think again. Harvard Bus Rev, 25 Oct 2011. https://hbr.org/2011/10/reorganizing-think-again.html

Asimov I (1950) I, Robot. Doubleday & Company, New York

Bean R (2016) Variety not volume is driving big-data-initiatives. MIT SMR Big Idea: Data & Analytics Blog, 28 Mar 2016. http://sloanreview.mit.edu/article/variety-not-volume-is-driving-big-data-initiatives/

Berman AE (2016) We are what we make: manufacturing's digital revolution is here. Posted on singularityHub, 18 May 2016, Singularity University, USA

Bidshahri R (2016) How AI will redefine love. Posted on singularityHub, 5 Aug 2016, Singularity University, USA

Biology Online (2017) www.biology-online.org/dictionary/Hybrid. Requested 19 Apr 2017

Blenko MW, Mankins MC, Rogers P (2010) The decision-driven organization. Harvard Bus Rev, June 2010. https://hbr.org/2010/06/the-decision-driven-organization

Brynjolfsson E, McAfee A (2014) The second machine age: work, progress, and prosperity in a time of brilliant technologies. W. W. Norton & Company

Deloitte (2016) Corporate culture threatens digital progress. 2016 Deloitte CIO Report. http://deloitte.wsj.com/cio/2016/08/02/corporate-culture-threatens-digital-progress/

Fan S (2016) Will letting AI make our decisions be the best decision we make? Posted on singularityHub, 11 Sept 2016, Singularity University, USA

Ganz J (2016) Pokemon Go is a glimpse of our augmented reality future. Posted on Singularity Hub (http://singularityhub.com/), 12 July 2016

Gratton L (2016) Technology and the "End of Management". FrontiersBlog, MIT SMR Frontiers, 27 July 2016

Hackman JR (2012) From causes to conditions in group research. J Organ Behav 33:428–444. doi:10.1002/job.1774

Hofstetter Y (2014) Sie wissen alles—Wie intelligente Maschinen in unser Leben eindringen und warum wir für unsere Freiheit kämpfen müssen. C. Bertelsmann Verlag

Illich I (1973) Tools for conviviality. Harper and Row, ISBN 0-06-080308-8, ISBN 0-06-012138-6

Jurdak R (2016) What if intelligent machines could learn from each other? The Conversation 2016. https://theconversation.com/what-if-intelligent-machines-could-learn-from-each-other-60102

Kane GC (2015) Are you ready for the certainty of the unknown? Big Idea: Social Business Interview, 10 Mar 2015, MIT SMR. http://sloanreview.mit.edu/article/are-you-ready-for-the-certainty-of-the-unknown/

Kaplan S, Orlikowski W (2014) Beyond forecasting: creating new strategic narratives. MIT SMR Magazine, Fall 2014 Research Feature, 16 Sept 2014. http://sloanreview.mit.edu/article/beyond-forecasting-creating-new-strategic-narratives/

Luhmann N (2012) Theory of society, vol. 1, translated by Rhodes Barrett. Stanford University Press

Mayer-Schönberger V, Cukier K (2013) Big Data—Die Revolution, die unser Leben verändern wird. Redline Verlag

Müller-Wirth M, Wefing H (2016) ICH. An article in the German weekly newspaper DIE ZEIT, 3 Nov 2016

NAP (2016) Art, design and science, engineering and medicine frontier collaborations: ideation, translation, realization: Seed Idea Group Summaries. The National Academies Keck Futures Initiative, The National Academies Press. ISBN 978-0-309-44347-0. doi:10.17226/23528

Oermann NO (2013) Tod eines Investmentbankers, Eine Sittengeschichte der Finanzbranche. Verlag Herder

Peters R (2015) Putting culture at the center of organizational change. 6 Aug 2015 in Organization, Change, & HR retrieved from http://flevy.com/blog/category/organization/Change.pdf

PWC   http://www.strategyand.pwc.com/media/file/Strategyand_Cultures-Role-in-Enabling-Organizational-Change.pdf

Ransbotham S (2016) In analytics, resolution must be accompanied by resolve. MIT SMR Blog

Ransbotham S, Kiron D, Prentice PK (2016) Beyond the hype: the hard work behind analytics success. MIT Sloan Manag Rev, Mar 2016. http://sloanreview.mit.edu/analytics2016

Reeves M, Zeng M, Venjara A (2015) Algorithms can make your organization self-tuning. BCG Perspect, 29 May 2015. https://www.bcgperspectives.com/content/commentary/business-unit-strategy-growth-reeves-algorithms-can-make-your-organization-self-tuning/

Rifkin J (2009) The empathetic civilization. Polity, Cambridge

Saenz A (2010) We live in a jungle of Artificial Intelligence that will spawn sentience. Posted on singularityHub, 10 Aug 2010, Singularity University, USA

Setz C (2014) der Digitale Adam. Die Zeit v 10 July 2014

Silvestro R (2016) Do you know what really drives your business's performance? MIT SMR Magazine: Summer 2016 Issue Research Feature 2 June 2016. http://sloanreview.mit.edu/article/do-you-know-what-really-drives-your-businesss-performance/

Smith MK (2001) 'Ivan Illich: deschooling, conviviality and the possibilities for informal education and lifelong learning', the encyclopedia of informal education. http://www.infed.org/thinkers/et-illic.htm

Stolterman E, Fors AC (2004) "Information Technology and the Good Life". Information systems research: relevant theory and informed practice. p 689. ISBN 1-4020-8094-8

Tapscott D, Tapscott A (2016) Blockchain revolution: how the technology behind bitcoin is changing money, business, and the world. Penguin Random House LLC

Wada K (2008) Robot therapy for elders affected by dementia. In: Gerontechnology. IEEE Eng Med Biol Mag 27(4, July–Aug 2008). doi:10.1109/MEMB.2008.919496

Westerman G et al (2011) Digital transformation: a roadmap for billion-dollar organization (Report). MIT Center for Digital Business and Capgemini Consulting

Chapter 7
Summary and Outlook

Abstract The Digital Transformation as a societal meta-development affects our organizations massively and forces them to act. The new conceptual framework for developing and improving an organization-facing these challenges of the digital transformation-with the (easy to memorize) metaphoric view of the subconscious mind and the conscious mind of organizations will organizations help to travel the bumpy road of digital transformation and disruption. Decision Making in this new "playing field" is in the focus of these considerations-concerning organization, tools, responsibilities, liabilities etc.. Society, politics and legislation are as well challenged by the digital transformation. Organizations may serve as role models for the societal developments when they succeed. The guiding principles in this book are intended to support this. The massive progress in AI and deep learning pushes organizations to a closer collaboration between man and machines–hopefully as man's tools and not as his dominators. To form and shape sensible hybrid intelligences and weave them in the organizations and their subconscious minds is a key challenge. Finally we must be aware of the singular qualities and abilities of the human to stay in the "driver's seat" on the road through this "stormy" digital transformation. Politics will have to speed up with regulations and legislation as well as adaption of macroeconomic structures, tax systems and shaping of the commons-the exponential technologies as AI, synthetic biology etc. "exploding" at the same time won't wait for regulation in favor of our societies. Politics and legislation are challenged to restore a culture of trust in the internet and its tools–for the benefit of organizations, economy and citizens. Maybe this will require the banning of anonymity as driver for criminal action, manipulation and hatred. There are better ways to enable whistle blowing where necessary.

Keywords Digital transformation · Society · Regulation · Decision-making · Legislation · Hybrid intelligencies · Trust · Anonymity in internet

The digital transformation is said to be an evolutionary step for mankind sometimes resulting in disruption—for example in entirely new products, services and business models—, but definitely resulting in paradigm shifts in society. These shifts result from our exponentially accelerating ability to create and process information, to communicate and to share sentiment, ideology and opinion. This is changing the perception and construction of reality and thus affecting and changing the roles of individuals, technology and organizations in the society.

Following Lee (2000) writing about the sociologist Niklas Luhmann I would like to add the following citation: "Social systems differentiate and evolve as they communicate in three separate dimensions: the social, temporal, and functional. The path of evolution results from a history of variation, selection, and restabilization within these dimensions. Communication, bit by bit, produces social structures that, recursively, produce future structures. Society is communication. Sociology, as the science of society, is communication about how different societal systems operate, communicate, evolve, and maintain their boundaries." Communication in society has been and still is massively changing with AI and hybrid intelligences becoming "part of the game". This is valid from a societal, interorganizational as well as an intraorganizational view and requires new perspectives on what is going on in order to be able to manage it and to master the challenges arising.

The framework of the subconscious mind of organizations and the concept of hybrid intelligences becoming part of our organizations and shaping them, might be of help from the organizational point of view and the respective challenging tasks ahead.

As concept and metaphoric view the "subconscious mind of organizations" serves as a relevant perspective when further developing our organizations in this phase of digital transformation. This framework shall help you to form and shape your organization actively in a conscious design process within the meanings of the guiding principles described in this book (which certainly can be further expanded and developed in the dynamic technological development we are in). The section titles in chapter 6 may also serve as guiding principles when forming hybrid intelligencies. This might be helpful for managers at all levels to thus come to a new holistic understanding of their organization in order to consider, experience and learn how their organizations work and how to develop their organization in this phase of transition.

This is especially relevant given the dynamic technological developments in the areas of perception, cognition, knowledge management and "Knowledge Discovery", decision-making, automation and robotics as well as Artificial Intelligence (AI) and blockchain. There clearly is a need to integrate them in our organizations wisely. The metaphor of "hybrid intelligences" becoming part of our organizations can help us to come to a close and targeted collaboration between man and machine (or better man and his tools) within our organizations and outside with other organizations in new business models and processes—with us as human beings trying "to stay in the driver's seat".

We have to find answers to the question of how to cope with the given uncertainty and the rising complexity induced by these ever accelerating processes and developments in technology and society. We are increasingly losing and missing the former dampening effects of slowness. This requires a "new thinking of organizations" and will force us to also change paradigms in economy and society as well. The golden calf of economic growth and the ways to measure, value and remunerate economic and societal activity as well as its outcome will have to be different—due to the digital transformation in many areas of life, due to Artificial Intelligence and the emerging changed "subconscious minds" of multiple, over-lapping and interacting organizations as well as societal groups—thus changing our societies as a whole.

In order to stay in the driver's seat we should be well aware of the qualities and abilities of humans that will be hard or impossible to emulate in technology as there are: experience based on perception with all senses, common sense, metacognition, morals, generalisation, imagination, dreaming, creativity, abstraction, metaphorism, compassion, empathy, emotion, solving dilemmas etc. Human relationships and respect as well as bonding are deeply human needs and are not ready to be digitalized-as are human learning and gathering experience which are grounded on human encounter.

In this context—also facing the development of chatbots generating comments and content with algorithms and their ability to thus manipulate opinion forming processes in social media—we will also have to discuss to abandon anonymity in the internet requiring authentification of users and responsible and liable persons for organizations. This is urgent when we want to keep this an ecosystem of com-munication and innovation for our organizations, providing trust as "bonding agent" of our societies, when we want to keep our societies safe and resilient, when we want to cope with cybercrime and when we want to keep something we call democracy. We have to fight to end up in a positive scenario, that in the end.

AI makes human intelligence more valuable than ever.

Epilogue

Let us—using the depth of considerations about technologies and organizations after reading this book as setup—finally have a look beyond our organizations, at the economy as a whole and at our societies.

The resource crisis of our planet is just as often a media issue as the—expected—development of economic growth. The paradigm of continuous growth is part of the "subconscious" of our Western society and therefore also of the global economy. From an evolutionary point of view this—as many scientists tell us—will lead us into the abyss. One can only appeal to those who are in charge and responsible, to rapidly and gradually reshape this part of the "society's subconscious mind", to change the way we measure and evaluate economic performance or to seek a more disruptive approach. Digital disruption and transformation is changing our society profoundly but we are not about to change how we evaluate human paid, unpaid and voluntary work and how we evaluate the work of machines and AI's to the best of social development, to the best of our societies. We should not wait for the total crisis as impetus. There is a high risk that this will lead us to uncontrolled changes and social upheavals.

Let me—as begun in the preface—end with a small fictional narrative close to driving—the car (still) being probably one of the best symbols of our Western civilization.

A group of international managers are—on the way to a solar energy project site—driving a road across the Arabian desert in highly sophisticated self-driving cars—the latest technology from the luxury segment. First, the route leads on highways and then it continues on typical sand roads. Over the radio, the cars of the convoy are interconnected. The passengers are talking about the bad economic data in the region, in Europe and the USA as well as the increasingly frequent natural disasters in the world. But the now available technologies of energy storage and energy transport make them optimistic about a recovering economic growth after the past seven "years of drought." And the carbon dioxide emissions now are finally also decreasing!

The weather forecast has predicted a medium sand storm. The navigation system can automatically choose the best route to avoid the central area of the sandstorm. Every passenger has a bottle of water with him. That is usually all

© Springer International Publishing AG 2017
W. Leodolter, *Digital Transformation Shaping the Subconscious Minds of Organizations*, DOI 10.1007/978-3-319-53618-7

they need. You do not get so thirsty with the air condition providing cool temperature and enough humidity in the car. However, the sandstorm is enormous.

Suddenly the navigation system no longer works. They can no longer receive satellite signals. Lately there were reported some disturbances with the satellite telephone resulting from too old equipment in the sky. That was due to the public budgets more concentrating on military infrastructure than on the maintenance of the civilian infrastructure. There were also recent reports about terrorist and cybercrime attacks with new technologies on sensible public space-based infrastructures. Due to recession in the last decade public infrastructure was rather neglected. But nobody expected those satellites and GPS to fail. The storm was heavy so that the street now could not be seen any more. The convoy stopped. There was no map and no compass in the car. There was some fuel left for keeping the air condition on. There was only one bottle of water for every passenger left … good luck.

The manufacturer's authorised representative in the EU is Springer
Nature Customer Service Centre GmbH, Europaplatz 3, 69115 Heidelberg,
Germany. If you have any concerns regarding our products, please
contact ProductSafety@springernature.com

Printed and bound by CPI Group (UK) Ltd, Croydon, CR0 4YY
29/04/2026
02099466-0003